T0146365

Detecting Conspiracy Theories on Social Media

Improving Machine Learning to Detect and
Understand Online Conspiracy Theories

WILLIAM MARCELLINO, TODD C. HELMUS, JOSHUA KERRIGAN,
HILARY REININGER, ROUSLAN I. KARIMOV,
REBECCA ANN LAWRENCE

Sponsored by Google's Jigsaw unit
Approved for public release; distribution unlimited

NATIONAL SECURITY RESEARCH DIVISION

For more information on this publication, visit www.rand.org/t/RRA676-1

Library of Congress Cataloging-in-Publication Data is available for this publication.
ISBN: 978-1-9774-0689-7

Published by the RAND Corporation, Santa Monica, Calif.
© 2021 RAND Corporation
RAND® is a registered trademark.

Cover: miakievy/Getty Images; Icons: Cecilia Escudero/iStock/Getty Images.

Support RAND
Make a tax-deductible charitable contribution at
www.rand.org/giving/contribute

www.rand.org

Preface

Conspiracy theories circulated online over social media contribute to a shift in public discourse away from facts and analysis and can contribute to direct public harm. Social media platforms are encountering a difficult technical and policy challenge as they try to mitigate harm from online conspiracy theory language. As part of an effort to confront emerging threats and incubate new technology to help create a safer world, Google's Jigsaw unit asked RAND Corporation researchers to conduct a modeling effort to improve machine-learning technology for detecting conspiracy theory language by using linguistic and rhetorical theory to boost performance. We also aimed to synthesize existing research on conspiracy theories with new insight from this improved modeling effort. In this report, we share insights from the effort and offer recommendations to mitigate harm and reduce the effects of conspiracy theories online.

The research reported here was completed in January 2021 and underwent security review with the sponsor and the Defense Office of Prepublication and Security Review before public release.

This research was sponsored by Google's Jigsaw unit and conducted within the International Security and Defense Policy (ISDP) Center of the RAND National Security Research Division (NSRD). NSRD conducts research and analysis for the Office of the Secretary of Defense, the U.S. Intelligence Community, U.S. State Department, allied foreign governments, and foundations.

For more information on the RAND ISDP Center, see www.rand.org/nsrd/isdp or contact the director (contact information is provided on the webpage).

Contents

Preface iii
Figures vii
Tables ix
Summary xi
Acknowledgments xix
Abbreviations xxi

CHAPTER ONE
Introduction: Detecting and Understanding Online Conspiracy
    Language 1
Research Approach 2
Hybrid Models: Dramatic Improvements in Performance Plus Insight 5

CHAPTER TWO
Making Sense of Conspiracy Theories 7
Literature Review 7
Text Analytics 14
Rich Descriptions of Conspiracy Theories 17
Key Findings on Conspiracy Theories Online 22

CHAPTER THREE
Modeling Conspiracy Theories: A Hybrid Approach 27
Data 27
Methodology 28
Model Performance: Hybrid Model Improves Performance 33
Key Insights from Our Modeling Effort 39

CHAPTER FOUR
Conclusion and Recommendations 41
Policy Recommendations for Mitigating the Spread of and Harm from
    Conspiracy Theories 42

APPENDIXES
A. Data and Methodology 45
B. Stance: Text Analysis and Machine Learning 59

References 73

Figures

3.1. Hybrid Model Overview 32
3.2. Topic Confusion Matrices Comparison Between Models 34
3.3. Confusion Matrices by Conspiratorial Qualities 36
A.1. Proxy Linguistic Stance Model Error 53
A.2. Hybrid Model Overview 54
A.3. Confusion Matrices by Conspiratorial Qualities 55
B.1. Aliens Language Category Stance Importance 69
B.2. Anti-Vaccination Language Category Stance Importance 70
B.3. COVID-19 Language Category Stance Importance 71
B.4. White Genocide Language Category Stance Importance 72

Tables

2.1.   Data Query Parameters 16
3.1.   Model Performance for Topics 37
3.2.   Model Performance for Conspiracy 37
A.1.   Data Query Parameters 46
B.1.   Definitions of Stance Variables and Categories 60

Summary

Conspiracy theories are an important part of what the RAND Corporation refers to as *Truth Decay*—a shift in public discourse away from facts and analysis caused by four interrelated drivers:

1. an increasing disagreement about facts and analytical interpretations of facts and data
2. a blurring of the line between opinion and fact
3. an increasing relative volume, and resulting influence, of opinion and personal experience over fact
4. a declining trust in formerly respected sources of factual information.

Conspiracy theories reflect both a move away from factual truth and declining trust in factual sources, replacing trustworthy information with untrustworthy information. Social media has made information, including conspiracy theories, easy to share across the global communications network.

Social media platforms are concerned about malicious or harmful uses of their services, and as part of their effort to combat harmful content on their platforms, Google's Jigsaw unit asked our RAND research team to help answer a difficult question:[1] *How can we better detect the spread of conspiracy theories at scale?* The scale of text on the internet is so vast that even large teams of humans can detect or flag only a fraction of harmful or malicious conspiracy theory language. Only machines can operate at that speed and scale. Jigsaw leaders fur-

---

[1] Jigsaw seeks to address technology threats and innovate for safer digital technologies.

ther understood that tackling this issue is more than just an engineering challenge. Because the spread of conspiracy theories is a sociocultural problem, they wanted more than a black box and asked whether we could advance machine-learning (ML) applications to provide additional insight: How do online conspiracies function linguistically and rhetorically?

The ability to detect a variety of conspiracy theories at scale while understanding their functional and persuasive features is an important step in addressing the problem through evidence-based interventions. Given not only the harm posed by existing conspiracy theories, but also the proliferation of new ones—for example, that anti-fascist activists in the Antifa movement started fires in Oregon in summer 2020—we feel this report is both timely and urgent.

Research Approach

Our research team mixed ML and qualitative research to better understand and detect online conspiracy talk by using the following methods:

- The first part of the study was a review of existing scholarly literature on conspiracy theories, followed by a text-mining analysis to try to understand how various conspiracies function rhetorically.
- The second part of the study was building improved ML models to detect conspiracy theories at scale.

Four Conspiracy Theories

For the research used in this report, we pulled data from Twitter that characterized four separate conspiracy theories about the existence of alien visitation, the danger of vaccinations, the origin of coronavirus disease 2019 (COVID-19), and the possibility of white genocide (WG). The alien visitation conspiracy theory offered a contrast to the others; it provides an example of an ideology that appears relatively benign.

Understanding How Conspiracy Theories Work

To better understand how conspiracy theories function, we first conducted a literature review to capture the state of knowledge on this topic. We then conducted a mixed-method analysis of online conspiracy theory language, using computer text-mining to detect patterns in our conspiracy theory data sets along with human qualitative analysis to make sense of how those patterns function persuasively.

For this effort, we used the stance analysis capabilities in RAND-Lex.[2] *Stance analysis* is a text-mining approach used to determine how speakers represent the world linguistically—the style and tone that point to the sociocultural part of language. One example is *certainty*: a writer could choose to use hedging language ("I think," "maybe," "it's possible that") or to use epistemic certainty markers ("we know," "it has been shown," "there is"). Those are representational choices that speakers make in attempting to achieve social effects (such as persuasion) within cultural contexts (such as genre and setting).

Improving ML Detection Through Hybrid Modeling

Second, we built an ML model that would detect a variety of conspiracy theories. We innovated by creating a hybrid model that combined word embedding (semantic content) with linguistic stance (rhetorical dimensions). ML has already made great progress in recognizing the semantic content of text—for example, automatically detecting whether an article is about sports, hobbies, or world events. Word embeddings using a deep neural network (DNN) are an example of a powerful way to classify documents (one that accounts for words as they appear in context) and thus do a very good job of capturing the semantic meaning of documents.[3]

To capture stance in our model, we operationalized a taxonomy of rhetorical functions of language originally developed at Carnegie

---

[2] RAND-Lex is RAND's proprietary text and social media analysis software platform. It is a scalable, cloud-based analytics suite with network analytics and visualizations, a variety of text-mining methods, and ML approaches.

[3] *Classifying documents* refers to assigning documents or text to a human-established set of classes. Also called *human-supervised learning*, this might, for example, mean inserting examples of threatening language, angry-but-not-threatening language, and neutral language into an ML model and then teaching the model to classify new documents into one of those classes.

Mellon University. We have used stance by itself in previous modeling efforts and gotten good results—for example, when detecting Russian interference in elections solely through rhetorical style.

This hybrid modeling effort is important for two reasons. First, although ML is getting better all the time at recognizing text content, recognizing rhetorical dimensions has been challenging. It is one thing to identify an anti-vaccination *topic*; it is a very different thing for a machine to interpret the *conspiratorial* dimension of anti-vaccination talk, and the latter angle is critical if we want to distinguish between talk that promotes conspiracy theories and talk that opposes or simply addresses them. The second reason that hybrid modeling is important is that DNN models, although powerful and useful, are also black boxes. Stance in this context has an interpretable representation and is not so high-dimensional that it cannot be looked at by humans. Thus, a hybrid model using DNN's semantic capability combined with stance would allow us to rank the importance of different rhetorical features and thus better understand how various conspiracy theories function rhetorically.

Hybrid Models: Dramatic Improvements in Performance Plus Insight

Our modeling effort was successful. We saw overall improvements in detecting conspiracy theory topics and dramatic improvements in discerning genuinely conspiratorial content within conspiracy topics. Furthermore, because we could output how the model weighted different stance features, we gained valuable insight regarding how various conspiracy theories function rhetorically. For example, we found that matters of public virtue, such as health and safety, were the most-important features that our model used for predicting anti-vaccination talk. Combined with our literature review, that sort of insight allowed us to develop possible interventions to help mitigate harm from online conspiracy talk.

Key Insights from Our Modeling Effort

We found that a hybrid approach to modeling conspiracy theory language worked well and offers several benefits over other current approaches. One important benefit is that a hybrid approach of stance and deep neural network word embeddings (e.g., Bidirectional Encoder Representations from Transformers [BERT]) appears to function as an out-of-the-box way to inductively capture genre features, obviating the need for specialized training for generic pretrained models, such as BERT. Another benefit is that adding stance allows us to create models that are more interpretable and to better understand the degree to which semantic content (the BERT portion) and the various stance features contribute to classification. This interpretability is critical for such tasks as dealing with conspiracy theory language, in which insight is as important as performance. Finally, hybrid modeling drastically reduced false positive rates, generally cutting them in half. We think this is very important from the perspective of social media platforms that wish to avoid flagging or moderating nonharmful content.

Policy Recommendations for Mitigating the Spread of and Harm from Conspiracy Theories

We found that our novel hybrid approach to ML could both improve performance and provide new insights regarding how online conspiracy theories function. By combining powerful existing ML approaches (BERT) with domain knowledge from linguistics and rhetorical studies (stance), we were able to advance practice specifically in the detection of conspiracy language with implications broadly for ML classification of documents that are marked more by sociocultural meaning than semantic content.

This innovation was the direct result of Google's Jigsaw unit framing the problem not simply as a technical challenge but as a sociocultural one that required a holistic approach. We think this sort of openness to improving ML through the creative use of insights from social science and domain experts is important as we confront the scale of difficulty in countering conspiracy theories specifically and Truth Decay more broadly. We hope that other social media platforms will follow suit and embrace creative approaches to sociocultural problems that go beyond purely technical solutions.

In addition to the practical output of an improved ML model to detect conspiracy theories, we also synthesized the model outputs of our effort with best practices derived from existing research literature. Understanding the rhetorical function of harmful conspiracy theories can inform evidence-based interventions that reduce adherence to and spread of these theories. We close out our report with recommendations for mitigating the spread and harm from online conspiracy theories.

Transparent and Empathetic Engagement with Conspiracists

The open nature of social media offers numerous opportunities to engage with conspiracy theorists. These engagements should not aggravate or provoke adherents: Instead of confrontation, it might be more effective to engage transparently with conspiracists and express sensitivity. Public health communicators recommend engagements that communicate in an open and evidence-informed way—creating safe spaces to encourage dialogue, fostering community partnerships, and countering misinformation with care. Validating the emotional concerns of participants, in particular, could encourage productive dialogue.

An additional technique beyond flagging specific conspiracy content is *facilitated dialogue*, in which a third party facilitates communication (either in person or apart) between conflict parties (Froude and Zanchelli, 2017). This approach could improve communication between authoritative communities (such as doctors or government leaders) and conspiracy communities. Facilitated dialogues could also be carried out at lower levels in the form of facilitated discussions that acknowledge fears and address feelings of existential threat for the participants.

Providing Corrections to Conspiracy-Related False News

One possible intervention for public health practitioners is to correct instances of misinformation using such tools as real-time corrections, crowdsourced fact-checking, and algorithmic tagging. In populations that hold preexisting conspiratorial views, the evidence for the effectiveness of corrections is mixed; however, results are consistently positive in studies investigating corrections of health-related misinformation in general populations.

Overall, the weight of the evidence appears to support such corrections. In addition, efforts to correct misperceptions in conspiracy-prone populations should follow the advice of public health practitioners and do so in the empathetic manner we have recommended. These efforts should be conducted in a manner that is transparent and sensitive to the concerns of conspiracy-prone audiences.

Engagement with Moderate Members of Conspiracy Groups

Conspiracists have their own experts on whom they lean to support and strengthen their views, and their reliance on these experts might limit the impact of formal outreach by public health professionals. Our review of the literature shows that one alternative approach could be to direct outreach toward moderate members of those groups who could, in turn, exert influence on the broader community. Commercial marketing programs use a similar approach when they engage social media influencers (or *brand ambassadors*), who can then credibly communicate advantages of a commercial brand to their own audiences on social media.[4] This approach is supported by academic research suggesting that people are more influenced by their social circles than mass communication (Guidry et al., 2015). For example, it might be possible to convey key messages to those who are only "vaccine hesitant," and these individuals might, in turn, relay such messages to those on anti-vaccination social media channels.[5] Similarly, religious or political leaders or political pundits who harbor moderate views could influence WG members.

---

[4] Influencer engagement programs have also been recommended as a strategy to counter violent extremism (Helmus and Bodine-Baron, 2017).

[5] Some have not yet decided to commit to the anti-vaccine cause, others opt for some but not all vaccines, and still others prefer administering vaccines in a more gradual schedule than the Centers for Disease Control and Prevention recommends.

Fears and Existential Threats

Underlying fears in the anti-vaccination and WG groups appear to powerfully motivate both groups. For anti-vaccination advocates, the fear rests on concerns regarding vaccine safety; for those concerned about WG, that fear rests on a belief in the (perceived) existential threat to the white race. To the extent that interventions can address such fears, they might be able to limit the potential societal harms caused by both groups. Efforts that target those who are vaccine hesitant, for example, could seek to understand concerns regarding vaccine safety and address those concerns by highlighting research on vaccine safety, the rigorous methods used in vaccine safety trials, or the alternative dangers that await those who are not vaccinated. For those concerned about WG, given the finding that some conspiracists are willing to engage in rational debate and that successful persuasion requires using the intended audience's values rather than the speaker's values (Marcellino, 2015), it might be more persuasive and effective to address claims that minorities will annihilate whites than to attempt to promote themes of racial equality.

Acknowledgments

We are grateful to numerous individuals and entities that supported the conduct of this research. In particular, we are grateful to Yasmin Green, Beth Goldberg, and Miranda Callahan at Google's Jigsaw unit for their willingness to support this work and for their close collaboration with us. We also gratefully acknowledge contributions from Luke Matthews of the RAND Corporation and Christine Chen of the Pardee RAND Graduate School. Finally, we would like to thank reviewers Christian Johnson of the RAND Corporation and Damian Ruck, formerly of Northeastern University and now Chief Researcher at Advai. Any errors in this report are the sole responsibility of the authors.

Abbreviations

BERT	Bidirectional Encoder Representations from Transformers
CMU	Carnegie Mellon University
COVID-19	coronavirus disease 2019
DNN	deep neural network
MCC	Matthews Correlation Coefficient
ML	machine learning
ReLU	Rectified Linear Unit
TF-IDF	Term Frequency-Inverse Document Frequency
WG	white genocide

CHAPTER ONE

Introduction: Detecting and Understanding Online Conspiracy Language

Conspiracy theories are an important part of what the RAND Corporation refers to as *Truth Decay*—a shift in public discourse away from facts and analysis caused by four interrelated drivers:

1. an increasing disagreement about facts and analytical interpretations of facts and data
2. a blurring of the line between opinion and fact
3. an increasing relative volume, and resulting influence, of opinion and personal experience over fact
4. a declining trust in formerly respected sources of factual information (Kavanagh and Rich, 2018, p. 3).

Conspiracy theories reflect both a move away from factual truth and declining trust in factual sources, replacing trustworthy information with untrustworthy information. Furthermore, the content of conspiracy theories can be inherently harmful and malicious. Anti-vaccination conspiracy theories threaten public health (André, 2003; Walker and Maltezou, 2014), and conspiracy theories that postulate a white genocide (WG) contribute to ethnically motivated violent extremism (Moses, 2019). Beyond broad public harm, conspiracy theories can contribute to acute individual harm, such as dehumanizing others along (perceived) racial and religious lines (Jolley, Meleady, and Douglas, 2020).

Social media has made information, including conspiracy theories, easy to share across the global communications network, and social media platforms are concerned about malicious or harmful uses of their services. As part of its effort to combat harmful content on its platforms, Google's Jigsaw unit asked our RAND research team to help answer a difficult question: *How can we better detect conspiracy theories at scale?*[1] It is not difficult for a human reader to see how an individual social media post or blog supports a conspiracy theory because humans bring lots of contextual knowledge to bear on text. But the scale of text on the internet is so vast that even large teams of humans can detect or flag only a fraction of harmful or malicious conspiracy theory language. Only machines can operate at that speed and scale. Furthermore, Jigsaw leaders understood that this is not just an engineering challenge. Because this is a sociocultural problem, they wanted more than a black box and asked whether we could advance machine-learning (ML) applications to also provide insight: *How do online conspiracies function linguistically and rhetorically?* The ability to detect a variety of conspiracy theories at scale while understanding their functional and persuasive features is an important step in addressing the problem through evidence-based interventions. Given not only the harm posed by existing conspiracy theories but also the proliferation of new theories—for example, that anti-fascist activists in the Antifa movement started fires in Oregon in summer 2020—we feel this report is both timely and urgent.[2]

Research Approach

Our team mixed ML and qualitative research to better understand and detect online conspiracy talk. Part of the study was a review of existing literature to try to understand how various conspiracies function rhetorically. The other part of the study was building improved ML models to detect conspiracy theories at scale.

---

[1] Jigsaw seeks to address technology threats and innovate for safer digital technologies.

[2] For example, see O'Sullivan and Toropin (2020).

Four Separate Conspiracy Theories

For the research used in this report, we pulled data from Twitter that characterized four separate conspiracy theories. We specifically selected conspiracy theories about the existence of alien visitation, the danger of vaccinations, the origin of coronavirus disease 2019 (COVID-19), and the possibility of WG. The alien visitation conspiracy theory offered a contrast to the others, providing an example of an ideology that appears relatively benign. We selected the topics of anti-vaccination and WG because of the risk of social and physical harm that could result from adoption and spread of these conspiracies. This reasoning also applies to COVID-19 conspiracies; the pandemic was at its height at the time of our study.

Using Text Analytics to Understand How Conspiracy Theories Work

We first conducted a literature review and analyzed the text that accompanied the conspiracy Tweets in our data set. Our goals were to capture the state of knowledge on conspiracy theories and to better understand the conspiracy theory content on Twitter. To accomplish the former, we conducted a structured review of scholarly research on conspiracy theories with a focus on interventions. To accomplish the latter, we used computer text-mining to detect patterns in our conspiracy theory data sets, along with human qualitative analysis to make sense of how those patterns function persuasively. For this effort, we used the stance analysis capabilities in RAND-Lex.[3] *Stance analysis* is a text-mining approach to understanding how speakers represent the world linguistically—the style and tone that points to the sociocultural part of language. One example is *certainty*: a writer could choose to use hedging language ("I think," "maybe," "it's possible that") or instead use epistemic certainty markers ("we know," "it has been shown," "there is"). Those are representational choices that speakers make in trying to achieve social effects (such as persuasion) within cultural con-

---

[3]   RAND-Lex is RAND's proprietary text and social media analysis software platform. It is a scalable, cloud-based analytics suite with network analytics and visualizations, a variety of text-mining methods, and ML approaches.

texts (such as genre and setting). The full list of 119 stance categories, along with definitions and examples, can be found in Appendix A.

Improving ML Detection Through Hybrid Modeling

Second, we sought to build an ML model that would detect a variety of conspiracy theories. In particular, we hoped to create an innovative hybrid model combining word embedding (semantic content) with linguistic stance (rhetorical dimensions). ML has already made great progress in recognizing the semantic content of text (for example, automatically detecting whether an article is about sports, hobbies, or world events). Word embeddings using a deep neural network (DNN) are an example of a powerful existing way to classify documents (one that accounts for words as they appear in context) and thus do a very good job of capturing the semantic meaning of documents.[4] The word-embedding model we chose to use, Bidirectional Encoder Representations from Transformers (BERT) (introduced by Google in 2018), is a powerful representation of language. BERT provided a good foundation for our attempt to boost performance through the addition of theoretical knowledge about how language functions rhetorically using stance. To capture stance in our model, we operationalized a taxonomy of rhetorical functions of language originally developed at Carnegie Mellon University (CMU). We have used stance by itself in previous modeling efforts and gotten good results—for example, detecting Russian interference in elections solely through rhetorical style.[5]

This hybrid modeling effort is important for two reasons. First, although ML is getting better all the time at recognizing text content, recognizing rhetorical dimensions has been challenging. It is one thing to identify an anti-vaccination *topic*; it is a very different thing for a machine to interpret the *conspiratorial* dimension of anti-vaccination talk, and the latter angle is critical if we want to distinguish between

---

[4] *Classifying documents* refers to assigning documents or text to a human-established set of classes. Also called *human-supervised learning*, this might, for example, mean inserting examples of threatening language, angry-but-not-threatening language, and neutral language into an ML model and then teaching the model to classify new documents into one of those classes.

[5] For more on the stance taxonomy, see Ringler, Klebanov, and Kaufer (2018). For prior use of stance in modeling, see Marcellino et al. (2020).

talk that promotes conspiracy theories and talk that opposes or simply addresses them. The second reason that hybrid modeling is important is that DNN models, although incredibly powerful and useful, are also black boxes. Stance in this context has an interpretable representation and is not so high-dimensional that it cannot be looked at by humans. Thus, a hybrid model using DNN's semantic capability combined with stance would allow us to rank how important different rhetorical features are in the model, and thus better understand how various conspiracy theories function rhetorically.

Hybrid Models: Dramatic Improvements in Performance Plus Insight

Our modeling effort was successful. We saw overall improvements in detecting conspiracy theory topics, and dramatic improvements in discerning genuinely conspiratorial content within conspiracy topics. We explain the modeling efforts in detail in Chapter Two. (A more technical explanation aimed at data scientists is provided in Appendix A.) Furthermore, because we could output how the model weighted different stance features, we gained valuable insight regarding how various conspiracy theories function rhetorically. For example, we found that matters of public virtue, such as health and safety,[6] were the most-important features that our model used for predicting anti-vaccination talk. Combined with our literature review, that sort of insight allowed us to develop possible interventions to help mitigate harm from online conspiracy talk. We discuss this effort in Chapter Three (with technical details provided in Appendix B). Finally, we summarize our efforts and offer both policy and technical recommendations in Chapter Four.

---

[6] For our purposes in this study, it is useful to distinguish between semantic and rhetorical dimensions to words and phrases. All lexical items or bundles have semantic (informational) content, and they can all have a rhetorical (pragmatic) function. For example, *health* and *public safety* are words or phrases with semantic content, pointing to specific concepts. But they can also be grouped in a rhetorical taxonomy under *public virtues*—things that are desirable in the public sphere—as opposed to *public vices*, such as *pandemics* or *racial injustice*.

CHAPTER TWO

Making Sense of Conspiracy Theories

In this chapter, we provide a review of scientific literature on conspiracy theories and social media. In addition, we provide a qualitative analysis of the content of the four key conspiracy groups studied in Chapter Three, which we have labeled alien visitation, anti-vaccination, COVID-19, and WG.

Literature Review

We first conducted a systematic review of the literature on conspiracy theories and social media as a way of placing the results of our quantitative analysis in proper context. *Systematic reviews* are review papers that use systematic methods to collect individual published studies and analyze and synthesize the findings of those papers. In this case, we searched available scientific databases using a common search query, applied specific entry criteria to guide the studies we ultimately reviewed, and enlisted a team of our analysts to systematically code the results of those studies. These methods are described in greater detail in Appendix A.

Ultimately, we found 108 peer-reviewed studies published between 2003 and the spring of 2020. By design, we focused on studies that specifically addressed the topics of conspiracy theories and social media: We did not review non-conspiracy-related disinformation or misinformation research. We also focused on studies that relied on the collection and analysis of original data.

The vast majority of these reports focused on analyzing social media data, with a minority of reports employing survey methods, modeling, and case-study analyses. Facebook and Twitter were the data sources used most often, and anti-vaccination conspiracy theories were the most common topic, with research into fake news and disinformation rising in popularity starting in 2012.

Key Themes Across the Literature

Our review of the existing literature on conspiracy theories helped us build a descriptive picture of conspiracy theory populations, such as their demographics and online characteristics. We also synthesized research on interventions across a variety of online conspiracies. This section summarizes our key findings.

Conspiracy Theories Common on Social Media

Conspiracy theories are commonly held in North America, with approximately one-quarter to one-third of surveyed populations expressing conspiracy-related opinions.[1] For example, one survey revealed that nearly one-third of Canadians reported uncertainty about a link between vaccines and autism or a belief that vaccines can hurt children (Greenberg, Dube, and Driedger, 2017). A U.S. survey indicated that 25 percent of respondents reported believing there is some truth to the conspiracy theory that COVID-19 was planned (Schaeffer, 2020).

Pro-conspiracy and anti-vaccine content are also commonly found on social media, although specific rates can vary across studies. Nugier, Limousi, and Lydié (2018) documented that 11 percent of vaccine-related websites had conspiratorial content; Song and Gruzd

---

[1] To further assess prevalence rates, we briefly examined literature outside our study's database of social media focused articles. For example, Mancosu, Vassallo, and Vezzoni (2017) confirmed general suspicions that conspiracy beliefs are widespread in Italy. Drawing on a national survey sample, they found that the percentages of acceptance of conspiracy theories regarding adoption of vaccines, the moon landing, and ChemTrails to be 24, 20, and 21, respectively. Nearly 40 percent reported believing in a recent conspiracy that the Italian pharmaceutical industry impeded testing of an alternative therapy for neurodegenerative disease. Uscinski et al. (2020) found that over 29 percent of U.S. survey respondents agree that the threat posed by COVID-19 is exaggerated, and more than 31 percent agree that the virus was intentionally created and spread.

(2017) found that 65 percent of vaccine-related YouTube videos had anti-vaccination content. Other studies documented a 50–50 mix of pro- and anti-vaccine content (Meleo-Erwin et al., 2017; Buchanan and Beckett; 2014). Finally, a recent study suggests that the rate of anti-vaccine content has been on the decline since 2014 (Gunaratne, Coomes, and Haghbayan, 2019).

Characteristics of Users and Populations

Several studies investigated various characteristics of pro-conspiracy users and populations. These studies found that women and parents were active in posting anti-vaccine content on social media sites (Tomeny, Vargo, and El-Toukhy, 2017; Hoffman et al., 2019). Mistrust in science is also correlated with conspiracy beliefs (Hoffman, et al., 2019), as is reliance on social media for health-related information (Featherstone, Bell, and Ruiz, 2019). Other studies looked at the association with education, income, political persuasion, and religion. These studies suggest that conspiracy acceptance is correlated with low education levels (Tomeny, Vargo, and El-Toukhy, 2017; Glenski, Weninger, and Volkova, 2018), political conservatism (Hornsey et al., 2020; Featherstone, Bell, and Ruiz, 2019), and religiosity (Landrum, Olshansky, and Richards, 2019).[2]

Linguistic Content (SM Content)

Other studies analyzed the linguistic aspects of conspiracy-related social media content. Most documented unique linguistic and rhetorical differences between pro- and anti-conspiracy groups. These studies suggest a tendency for conspiratorial populations to manipulate scientific evidence, make use of emotional appeals, and manifest beliefs in more than one conspiracy theory.

Nugier, Limousi, and Lydié (2018) analyzed over 3,000 anti-vaccination websites and found six unique rhetorical strategies that anti-vaccination advocates used in arguing their points:

---

[2] Several studies discussed characteristics of populations that were less prevalent within our literature sample. Narayan and Preljevic (2017) examined characteristics of individuals who converted from advocating anti-vaccination views to advocating pro-vaccination views, finding that some individuals converted because of exposure to illness that could be prevented by vaccines. Bessi, Petroni, et al. (2016) reported that Facebook users who posted about conspiracy theories had higher levels of emotional stability.

1. manipulating scientific evidence
2. appealing to emotions
3. referencing conspiracy theories unrelated to vaccination issues
4. claiming that vaccines are unnatural practices
5. negative benefit-risk ratio
6. arguments about freedom of choice.

Numerous other studies validate these findings, especially in respect to manipulating scientific evidence (Takaoka, 2019; Ekram et al., 2019; Marcon, Murdoch, and Caulfield, 2017), appealing to emotions (Greenberg, Dube, and Driedger, 2017; Guidry et al., 2015; Mocanu et al., 2015; Brugnoli et al., 2019) and using broader conspiracy theories (Sommariva et al., 2018; Arif et al., 2018; Bhattacharjee, Srijith, and Desarkar, 2019; Gandhi, Patel, and Zhan, 2020; Hornsey et al., 2020; Mahajan et al., 2019; Penţa and Băban, 2014; Smith and Graham, 2019).

Engagement
Nine studies examined engagement (referring to the number of likes, shares, and comments) with conspiracy content on social media sites. These studies showed high rates of engagement for conspiracy content. Studies have shown for example, higher rates of engagement for anti-HPV vaccine Instagram posts than for pro-vaccine posts (Kearney et al., 2019), for misleading Facebook posts about the Zika virus than for accurate posts (Sharma et al., 2017) and for rumor-based online news articles about Zika than for verified news articles (Sommariva et al., 2018). Another study found more dislikes than likes for pro-vaccine YouTube videos (Song and Gruzd, 2017).[3]

---

[3] Other studies examining tone of engagement revealed mixed results. Cuesta-Cambra, Martínez-Martínez, and Niño-González (2019) found that there were no differences in emotional responses and engagement between anti-vaccine and pro-vaccine sites. In contrast, Briones et al. (2012) found that social media users preferred to engage with negative content: Negative videos with anti-vaccine messages were liked more than positive or ambiguous ones.

Networks, Info Flow, and Consumption Patterns

Twenty studies examined how social media networks related to conspiracy content. Studies found that conspiracy-minded social media users consume and interact with content within like-minded echo chambers while being adept at spreading conspiracy content throughout their social networks.

Studies found that conspiracy-minded social media users enter increasingly into echo chambers that serve to reinforce their beliefs. Bessi, Coletto, et al. (2015) found that Facebook users were more engaged with conspiracy content when it came from friends who had similar consumption patterns. Findings also suggest that polarized social media users are more likely to create small groups and reinforce misinformation (Wood, 2018). In addition, these echo chambers reinforce individual beliefs and help cement in-group relationships while leading to fewer interactions with non-conspiracists (Brugnoli et al., 2019; Bessi, Petroni, et al., 2016).

Conspiracy communities spread more information throughout networks than non-conspiracy populations. Glenski, Weninger, and Volkova (2018) found that the majority of misinformation spread within a Twitter network came from a small group of highly active users. Bessi, Zollo, Del Vicario, Scala, Caldarelli, et al. (2015) found that conspiracy users on Facebook tried to diffuse conspiracy news throughout the network while scientific populations preferred to stay within their own echo chambers. Similarly, Lutkenhaus, Jansz, and Bouman (2019) found that Twitter posts from anti-vaccine conspiracy theorists spread to other communities more than did scientifically valid health information from pro-vaccine users.

Interventions

This literature suggests mixed results for the effectiveness of corrections, and it highlights that such corrections might be more effective if they were presented with a more caring and humble tone that encouraged dialogue. We also briefly review interventions that might protect non-conspiracy populations from believing conspiracy-related misinformation.

Studies examining the effectiveness of correcting misinformation on social media yield mixed results with pro-conspiracy populations. Interventions designed to correct misinformation or conspiracies in online communities is challenging because strong adherents to conspiracies are strongly resilient against information that contradicts personal opinions. Such was the conclusion of a study by Giese et al. (2020), which exposed research participants to different pieces information about flu vaccinations that were consistent and inconsistent with participants' preexisting attitudes.[4] Another study seemed to demonstrate that polarized conspiracists who are exposed to contrasting or teasing debunking narratives actually increase their engagement with unsubstantiated rumors (Bessi, Caldarelli, et al., 2014).

But other studies suggest results that are more positive. Bode and Vraga (2018) documented a successful intervention: Algorithm-recommended or friend-recommended corrective news stories limited misperceptions about the Zika virus for conspiracy theorists.[5] In an earlier study, the same authors showed that corrections of health misinformation were actually more effective in audiences with high levels of misperceptions (Vraga and Bode, 2017).

Transparency and Sensitivity in Responding

Recent research suggests that corrections might be most effective when they are communicated in a transparent and sensitive manner. Gesser-Edelsburg et al. (2018) documented that both pro-vaccine and vaccine-hesitant participants preferred transparent corrections that addressed their concerns, including their emotional concerns. Public

---

[4] Likewise, Zollo et al. (2017) found that debunking information was ineffective among social media users and did not change the rate of user engagement with conspiracy content.

[5] Two other studies had successful results, but their impact is uncertain because of study design. Porreca, Scozzari, and Di Nicola (2020) studied YouTube video topics before and after Italy passed a vaccination law. The study showed that a vaccination campaign promoted by medical doctors pushed the sentiment in favor of vaccines. However, this is not a robust measure, and the study is confounded. It is unclear whether the sentiment shift is a result of the legal change or the doctors' campaign. In another study, Brainard and Hunter (2020) used an agent-based model to test two different types of interventions for countering misinformation that could worsen disease outbreaks. In the first approach, the authors increased the proportion of shared information that offered "good advice" that encouraged protective behaviors. In the second, they modeled efforts to "immunize" individuals so that they do not respond to or share bad advice. The model showed that both approaches reduce misinformation, though the interventions obviously need to be tested in human populations.

health communicators seem to agree and argue that corrections should done in ways that create safe spaces to encourage dialogue, foster community partnerships, and counter misinformation with care (Steffens et al., 2019), and that ad hominem attacks against vaccine skeptics are unlikely to persuade (Gallagher and Lawrence, 2020). Likewise, Zollo et al. (2017) highlighted that corrections should promote "a culture of humility" that demolishes "walls and barriers between tribes, could represent a first step to contrast misinformation."

Preventing the Spread of Conspiracy Theories

Other studies investigate methods to protect general non-conspiracy-minded audiences from the effects of misinformation. Although such studies were not picked up in the conspiracy-focused literature search, they are nonetheless worth a brief highlight.

First, corrections can mitigate the effects of misinformation. This was the conclusion of Walter et al. (2020), who examined 24 social media interventions (e.g., real-time corrections, crowdsourced fact-checking, and algorithmic tagging) designed to correct health-related misinformation.

Second, generalized warnings might help social media users discern accurate information from false information. Clayton et al. (2019) demonstrated that providing consumers with a general warning that subsequent content might contain false or misleading information increases the likelihood that they see fake headlines as less accurate. Clayton and her colleagues also documented the effectiveness of "disputed" or "rated false" tags.

It also might be possible to inoculate audiences against the effects of misinformation. Inoculation interventions work by pairing a warning message about misinformation with a weakened example and then offering guidance on how to refute the message. Studies have shown that inoculation procedures effectively induce resistance to conspiracy theories, extremist propaganda, and climate change misinformation (Cook, Lewandowsky, and Ecker, 2017; Braddock, 2019; Banas and Miller, 2013; van der Linden et al., 2017).

Finally, online media literacy interventions might be effective. Most media literacy interventions consist of holistic education programming, but research is starting to suggest that social media–based media literacy lessons can be effective. Guess et al., (2020) documented that Facebook-based media literacy content led to significant and persistent effects in the ability of participants to discriminate between mainstream and false news headlines. A recent RAND study (Helmus, Marrone, and Posard, 2020) also documented that a short media literacy video reduced the degree to which right-leaning social media users engaged with Russia propaganda memes.

Text Analytics

In addition to our modeling effort to identify conspiracy theory language online, we combined the ML model output with qualitative analysis of conspiracy theory language. The goal was to use *feature importance* (highest-ranked rhetorical features in the model) with analysis of text samples rich in those features to both describe and understand how conspiracy theory language works online. It is one thing to see that that highest-rated feature in our anti-vaccination language model is the language of public sphere good; it is another thing to understand how anti-vaccination proponents express care and concern for public health and the safety of their children.

Data

Social platforms place various limitations on the formatting of their content (e.g., Twitter's 280-character limit on the length of individual posts). This shapes the type of speech that one is likely to encounter on each platform. To ensure that modeling results account for this variation, we collected data from multiple sources. Our data collection was conducted through the social media tracking company Brandwatch.[6] This gave us access to archived content from multiple social media platforms, which we culled using queries in Boolean format and by filtering results in terms of location, time, language, and other criteria. For the purposes of this study, we queried for English only. Social media

---

[6]   Brandwatch (undated) was formerly named Crimson Hexagon.

sources were Twitter, Reddit, and a large selection of online forums and blogs.[7] We also used one-off sources, such as the transcript of the "Plandemic" viral video (2020).[8]

In consultation with Jigsaw, we selected four specific conspiracy theory topics: alien visitation, anti-vaccination content, COVID-19 origins, and WG.[9] Our social media queries were iterative; prior rounds informed subsequent rounds. Table 2.1 lists the search parameters and the final numbers of documents (tweets, blog posts, comments) for each of the four queries.

Analysis Method

We conducted a mixed method of text analysis: statistical reports of stance features by conspiracy theory, followed by human reading of feature-rich samples. This kind of analysis combines machine distant-reading for patterns in the data with human close-reading for meaning. To prevent harm and guard the privacy of those from whom we collected data, we do not use direct quotes. Instead, we paraphrase several quotes together to represent the language expressed by multiple users. These paraphrases appear in single quotes ('), and although multiple such paraphrases can appear in a specific example, the single quotes distinguish between different speakers and selections. Also, because this analysis featured human qualitative analysis at the level of individual posts, we were able to distinguish between concurring and dis-

---

[7]   Brandwatch collects data from over 1 million blogs, and examples include WordPress, Medium, Blogger, Typepad, TMZ, IGN, Engaged, Business Wire, Mashable, Techcrunch, Kottke, Business Insider, Gizmodo, IMDB, LifeHacker, The Verge, Hardwarezone, and TechRadar. Our search covered over a thousand online forums—for example, Yahoo Answers, Mumsnet, MyFitnessPal, Psychology Today, AVforums, Stack Overflow, Goodreads, Investopedia, GameSpot, FlyerTalk, Tianya, Naver, MacRumors, MoneySaving Expert, Market Watch, GlassDoor, The Student Room, and Steam Community.

[8]   We specifically included the Plandemic transcript because the words from it were so widely recirculated on social media. We used sources distinct from social media, such as Plandemic, because similar rhetoric often makes its way eventually into the social media sphere as talking points. For example, Plandemic is referenced directly in 6 percent of social media comments in this data set.

[9]   WG refers to the idea that genocidal campaigns against ethnically white populations are occurring globally.

Table 2.1
Data Query Parameters

Example Search Terms	Period Covered	Documents

Alien visitations
    aliens, ufo, extraterrestrials,
    visitation, roswell, tunguska, crop
    circles, oumuamua, annunaki, secret,
    government
	January 1–March 31, 2020	~30,000

Anti-vaccine myths
    vaccine, immunization, safety, harm,
    thimerosal, adjuvant, big pharma,
    cover-up, autism, infertility
	January 1–March 31, 2020	~160,000

COVID-19
    coronavirus, covid, lab, secret,
    government, bioweapon, man-
    made, 5g, gates, pirbright, deep
    state
	January 1–April 16, 2020	~60,000

White genocide white genocide, white rights	January 1–March 31, 2020	~50,000

senting perspectives in posts. For example, a response to a previous post supporting a given conspiracy theory might include a sarcastic rejoinder or insult, making it clear to a human reader that the response was on the same topic but opposed in perspective.

Stance analysis is both *quantitative* (statistical frequencies and distributions of language categories) and *qualitative* (rich descriptions of attitudes and beliefs inferred from the language categories). In the next section, we provide details about our qualitative analysis of the statistical results.

Rich Descriptions of Conspiracy Theories

Here, we provide key insights from each conspiracy community. Additional findings about how our model distinguishes these conspiracy theories from normal discourse are discussed in Appendix B.

Alien Conspiracy Community

Alien conspiracy speech was linguistically distinctive by its wide variety of stance features with small effect sizes, such as spatial relations ('travel to'), social closeness ('us'), uncertainty ('it's possible'), authority sources ('the government'), contingent reasoning ('could be'), and looking back ('long ago'). Using stances relating to spatial relations, social closeness, uncertainty, and looking back makes sense when suggesting the possibility that aliens have traveled to visit Earth, especially if this is framed as having happened in an ancient past. Unlike the anti-vaccination and WG conspiracies, there was little talk from detractors opposing any theories. Another feature of this data set was the expression of clearly developed conspiracies about government cover-ups of alien visitation. Comments suggest there is a large distrust of the government and what the government is telling the American public ('secret space program,' 'it's classified'). Personal roles figured in talk about aliens ('alien leader,' 'reptilian aliens'). There were more one-off comments and invitations to watch documentary videos or lectures on aliens than there were arguments with detractors. Regardless of the truth of this conspiracy theory, it seems much more innocuous than the other conspiracies we studied, with no antisocial component or direct threat to public health.

Anti-Vaccination Conspiracy Community

The anti-vaccination conspiracy community had the most distinctive stance profile of the four conspiracy groups (see Appendix A for details). Anti-vaccinators expressed distrust for vaccines; said they felt bullied; and expressed negative emotions, such as fear, anger, and frustration.

Distrust of Vaccine Safety

Anti-vaccinators expressed distrust of vaccines—specifically, the safety of vaccines, vaccine safety tests, and medical authorities who recommend vaccines despite what anti-vaccinators see as evidence for vaccine-related injuries. The most distinctive rhetorical feature in anti-vaccinators' online talk was *public virtue* speech (such as references to justice, fairness, happiness, health). Anti-vaccinators were particularly concerned with the safety of vaccines ('vaccine safety,' 'safe vaccines').

Discussion of *personal roles* was also very high; members of this group populated their talk with mothers, parents, and doctors. In practice, anti-vaccinators are discursively parents expressing deep concern for the safety of their children in contrast to distrusted medical authorities.

Comments using *public vice* speech (such as references to injustice, unfairness, unhappiness) cited conflicts of interest with vaccine safety validation tests, claiming, 'Big Pharma conducted those safety trials.' Because of these conflicts of interest, anti-vaccinators concluded that vaccine safety results are untrustworthy. Concerns about vaccine safety were also expressed in negative comments about vaccines being toxic ('get rid of toxic vaccines!' 'vaccines are TOXIC') or linked to autism, cancers, or attention deficit hyperactivity disorder. Anti-vaccinators made appeals to sympathetic medical authorities, celebrities, and pundits supporting anti-vaccination sentiments, making particular use of reporting stances in discussing specific anti-vaccination doctors and their lectures on vaccine injuries.

Bullying and Fear

Pro-vaccinator participants responding in these conversations often used derisive comments ('yaddah yaddah, vaccine injuries, yaddah, yaddah,' 'fake vaccine injuries'). So, in addition to expressing distrust of vaccine safety, anti-vaccinators expressed anger over feeling bullied ('bullying a mother of a vaccine injured kid?!' 'abusing parents of vaccine-injured children'). Public vice language was featured in heated arguments between anti-vaccinators ('the evidence is in vaccine-injured kids,' 'fraud vaccines') and pro-vaccinators. Anti-vaccinators also expressed fear of vaccine injuries and anger or frustration over vaccine injuries being ignored ('vaccines do harm!' 'vaccine injuries are no laughing matter'). Many negative and angry comments apparently stemmed from anti-vaccinators not feeling validated ('quit ignoring vaccine injured kids!').

In sum, anti-vaccinators talked about vaccines as unsafe, expressed frustration at dismissed claims of vaccine injuries, and, as they see it, being bullied for not wanting to put their children in harm's way. They expressed distrust of vaccine validation studies, articulated a conflict of interest with past validation studies, and said that vaccines cause injuries or other diseases. They reported feeling bullied and seemed emotional, mostly expressing distrust, anger, fear, and frustration over vaccine safety and ignored vaccine injuries. In terms of argument, anti-vaccinators used selective scientific authorities, doctors, and celebrity endorsements to dispute claims of pro-vaccinators.

COVID-19 Conspiracy Community

COVID-19 conspiracists principally discussed the origins of the virus and expressed distrust over mainstream sources of information. In this data sample, the discussion also focused on distrust of authority sources and speculation around shadowy forces presumably behind the virus's origin.

COVID-19 Origins

In our data set, COVID-19 conspiracists focused on the origins of the virus, not its spread, symptoms, or prevention measures. The most-prominent origin theories were: 'man-made' (the most repeated phrase of this community), 'Chinese bioweapon,' 'a product of the Deep State,' and the idea that COVID-19 was a cover-up for 'radiation illnesses' from 5G networks. Public virtue language was mostly through calls for 'truth' about COVID-19 origins. COVID-19 origins were marked by uncertainty language—hedging devices such as 'I'm not sure but' or 'maybe it.' This stance of uncertainty could be a marker of a novelty that helps identify the emergent quality of this conspiracy.

Unlike the anti-vaccination group, there were few opponents arguing against this group. Possible explanations are that the COVID-19 conspiracy group is new and would-be detractors are unaware of it, the actual pandemic demands so much attention that detractors cannot attend to conspiracy claims, or the claims of COVID-19 conspiracists do not pose societal harm and so detractors are not motivated to engage with it.

Distrust of Authority

These conspiracists also distrust authority and believe that those who produce the news are lying to them. Public vice speech was above baseline levels and frequently referenced fake news and what to blame for COVID-19 ('fake news,' 'science experiment poisoning humanity,' 'corrupt disgraceful leaders'). Conspiracists also expressed negativity and frustration as they placed blame ('5G Hoax,' 'China messed up,' 'Virus Hoax and 5G Syndrome,' 'Deep State WMD,' 'you're full of it'). The variety of those being blamed suggests that this group has not come to a consensus on COVID-19's origins. As a result of distrusting the news, conspiracists looked for other information sources and used reporting language as they shared the information found ('READ,' 'link,' 'download it all, it's a good read'). Personal roles were also important; this group expressed trust for some authorities ('Report from Dr. Jane Doe'), but—similar to anti-vaccinators—expressed distrust of other authorities ('Dr. Depopulation,' 'The Terminator').

Overall, the key features of the COVID-19 conspiracy group are that its members are most concerned with the (perceived) malicious origins of COVID-19 and that they distrust news sources, instead sharing alternative sources of information. The group is new and has not yet settled on the pandemic's origins; several sources and theories are being shared, and few detractors are commenting. The group does not discuss public health dimensions of pandemic (e.g., the risk of infection or mortality rates). They also do not advocate against such public health measures as physical distancing or wearing a mask in public. The greatest risk from this community might be threats to 5G cell phone towers, although actual attacks on such towers are rare.

White Genocide Conspiracy Community

WG conspiracists articulated a perceived existential threat to their racial group, responding with racist, antisocial language. They explicitly used an adversarial "us versus them" conceptual framework between white people and nonwhites and, to some extent, between WG believers and deniers. However, this group also engaged in a robust exchange between proponents and detractors in which conversational and various lines of reasoning seemed to be in use. This dialogic exchange could point to a potential line of intervention with this group.

Existential Threat

WG believers talked about an existential threat to the white race ('replacement theory,' 'murdering white people,' 'look at this classroom in Norway. Replacement!'), which might motivate the extreme speech in which this group engaged. Of the four conspiracy theory data sets we analyzed, the WG group was qualitatively the most antisocial, featuring name-calling ('you left retards'), negativity ('you suck at this'), and racist comments. From a stance perspective, WG speech was marked by the highest public vice speech of any group ('racist,' 'genocide,' 'Nazi'). This is mixed with concrete properties (i.e., black, white, brown); WG believers talk about the world in terms of skin color. Interestingly, their language was relatively low in references to concrete objects (physical objects in the real world)—in a sense, they linguistically construct a very abstract social world where "we" are threatened by "them," visible in the personal roles that mark their speech ('Jews,' 'Rabbis,' 'Christians,' 'white people').

Control Over Women's Bodies

We found that WG discourse was marked by a patriarchal concern over women's sexual activity. Many of the documents richest in concrete properties and personal roles were about 'white women' with 'black men.' WG conspiracists worried about white women going to Africa and having sex with African men, argued that '90% of white women that have babies with black men end up raising the child on their own,' and complained about diversity media that show 'a black man, with a white woman. You truly hate white people. This is white genocide.' While this concern dovetails with WG conspiracists' (perceived) existential fear of being racially erased, we note that the concern was solely around women—we did not see similar talk about 'white men' with 'black women.'

Argumentation Characteristics and Styles

The WG group frequently referenced political and historical figures and issues along with such contemporary topics as European political parties, Marxism, and historical events. Some WG group members referenced historically obscure figures who wrote or are purported to have written about diminishing the number of white people. There was also some mention of President Donald Trump and his slogan "Make America Great Again," but this was a minor theme.

Like anti-vaccinators, the WG community had an active exchange with detractors, to whom WG believers responded with a variety of argumentation styles. Conspiracists showed a mix of argument styles, such as hateful ('Jews hate Christianity'), simplistic ('brown people having more kids is white genocide!'), and insular or difficult to follow ('preventing slavery = white genocide'). Detractors expressed frustration ('what is wrong with you?!'), reasoned ('you can't argue in good faith with WG'), or identified humor ('you realize that's a joke, right?') in their responses. A small amount of public virtue speech present showed that conversational niceties ('take a look at this,' 'good one, thanks bro') were also exchanged in this otherwise heated conversation. Overall, the exchange of comments between WG believers and deniers seemed much more substantive and engaged than the one-sided comments typical in the anti-vaccination group. This willingness to argue could mean that interventions using reasoned argument might be useful in opposing WG conspiracy theories.

In all, the WG community was distinctive for its existential fear, high public vice speech, hate-based name-calling, and engaged dialogue. Given the finding from our literature review that it is possible to engage moderate subgroups within conspiracy-holding communities, there could be opportunities for engagement—the caveat to that being the deeply antisocial and strongly racist idea that "they" are a threat to "us" that forms the foundation of WG conspiracies.

Key Findings on Conspiracy Theories Online

We offer the following key findings, based on our literature review and our detailed analysis of online conspiracy theory language:

Conspiracy Content Popularity and Online Echo Chambers

We found that conspiracy beliefs are commonly held, and evidence suggests that more than one-quarter of adults in North America believe in one or more conspiracies. Conspiratorial content on social media is likewise common; anti-vaccine content, for example, is present on 11–65 percent of vaccine-related websites, YouTube videos, or social media posts. Pro-conspiracy theorists also find themselves wading deeper into social media–based echo chambers with decreasing exposure to non-conspiracy viewpoints. These echo chambers contribute to a deepening polarization of viewpoints, and the posts disseminated within such echo chambers can reach and influence the broader internet.

Detriments of Bullying

Analysis of anti-vaccination social media content suggests that pro-vaccinators routinely confront anti-vaccinators with condemnation, which could lead to an angry and fear-driven response on the part of anti-vaccinators. Hostile engagement risks further inflaming the opinions of anti-vaccinators. For example, Bessi, Caldarelli, et al. (2014) showed that exposure to debunking narratives that used a teasing tone led the most-polarized conspiracists to actually increase their subsequent interactions with unsubstantiated rumors. Such negative and unintended effects of persuasion campaigns are not uncommon and are referred to as a *boomerang effect* (Byrne and Hart, 2009).

Conspiracy Theory Qualities

The conspiracy theories we examined in this study were marked by qualitative variety. At the farthest end of this spectrum, the alien conspiracy discourse we studied was not marked by harmful or antisocial sentiment, although it is marked by distrust of the U.S. government. Anti-vaccine conspiracy talk poses overt threats to public health, directly opposing an important public health effort. In a very different way, the blatant racist views held and propagated by WG conspiracists are directly threatening and dehumanizing toward blacks, Jews, and other ethnic groups; such views also align with violent extremist ethnonationalism. In contrast, the COVID-19 conspiracy talk was not marked by direct opposition to public health efforts or overt antisociality, but conspiratorial ideas about the origins of the virus might enable other harms (such as anti-Chinese sentiment or distrust of vaccination efforts).

Reliance on Authority Figures

A common thread among all the conspiracy groups was distrust of conventional authority figures. However, each group could point to its own in-group authorities as sources of inspiration and knowledge. Anti-vaccinators express distrust of medical authorities while high-lighting opinions of sympathetic medical authorities, celebrities, and pundits. COVID-19 conspiracists distrust authority and mainstream media news, but they will still highlight material from agreeable medical authorities. Even the WG group made selective use of authority figures. This analysis suggests that intervention efforts need to be careful in the use of authority figures to counter conspiracist groups.

"Us Versus Them" Language

Although antisocial discourse involves the creation of social others—an *in-group* and *out-group* paradigm—not all language markers for "us" and "them" are inherently antisocial. RAND-Lex performs stance coding for social distancing and social closeness language that map onto that paradigm, which was most present in our alien visitation data set. In that case, talking about how '*they* traveled across the galaxy to visit *us*' requires "us" and "them" distinctions—but, in this case, the distinction was prosocial: Conspiracy theorists generally talked about aliens with positivity and excitement. A truly antisocial "us versus them" paradigm, such as that through which the WG conspiracists work—is a more complex set of rhetorical moves than simply "us versus them."

Anti-Vaccinators and Vaccine Safety

Various conspiracist communities carry their own concerns that help drive and motivate their views. In this study, anti-vaccinators primarily focused on vaccine safety. In the data we analyzed, members of this group worried about the safety of vaccines and the accuracy and legitimacy of vaccine safety tests. These views are wrapped up in a broader concern for public safety in general and for children specifically. Engagement efforts that empathically address this community's concerns about vaccine safety might mitigate the group's harm to public health.

White Genocide Advocates, Racial Annihilation, and Control of Women's Bodies

WG believers articulate a perceived existential fear that the white race is facing an existential threat from a variety of ethnic groups that conspiracists describe as black and brown. We also note a focus on the control of women's bodies—the idea of white women having sex with nonwhite men is another existential fear. Although these fears are deeply held, WG adherents represented in our data were willing to engage with detractors; thus, it might be possible to open dialogue and engage them.

CHAPTER THREE

Modeling Conspiracy Theories: A Hybrid Approach

Understanding and detecting conspiracy discourse in social media is a nontrivial task. Identifying a conspiracy requires context outside any one document, borrowing most heavily from cultural awareness, which can drift quite rapidly in social media. To make matters more difficult, conspiracy discourse can exist across several avenues of communication simultaneously. Twitter is a good example: Tweets might be accompanied by videos, images, and emoji symbols, all of which help communicate conspiracy discourse presented in text form. For this initial attempt at modeling conspiracy discourse, we bounded our attempt at text from social media. In the following sections, we share two hybrid model-building projects: one that aimed to identify social media posts addressing our four conspiracy topics (alien visitation, anti-vaccination, COVID-19 origins, and WG) and a second that aimed to identify whether a given text actually promotes the conspiracy.

Data

The kind of human-supervised ML that we conducted requires large *training data sets*: that is, many examples of the things a programmer is trying to teach the machine to classify. In consultation with Jigsaw, we selected four specific conspiracy theory topics to model: alien visitation, anti-vaccination content, COVID-19 origins, and WG. We collected 150,000 samples of social media material across these four types of conspiracy language along with a baseline sample of "normal" non-conspiracy talk, for five sets of 30,000 text samples each. To account for differences across various social media platforms, we collected data from multiple sources: Twitter, Reddit, a variety of online forums and blogs, and some single sources (such as the transcript of the "Plandemic" video [2020]).

We collected conspiracy talk through the social media tracking company Brandwatch (undated). This gave us access to archived content from multiple social media platforms, which we culled using queries in Boolean format and by filtering results in terms of location, time, language, and other criteria. This collection process was iterative; we implemented multiple rounds of constructing a query, getting a sample of data, and hand-inspecting samples from the data until our query captured talk that focused topically on our various selected conspiracy theories. Because our study had a U.S. focus (and because stance is language-specific), we collected English-language data only.

Our collection of baseline social media talk was an existing general corpus of tweets randomly sampled across 2015 via Gnip,[1] stratified to cover each month; weekdays and weekends; and morning, afternoon, evening, and late-night periods. The baseline we used is only Twitter data, and our training data set includes other talk (e.g., Reddit threads), but we were more concerned with a broad baseline of topics (and, thus, ways of speaking) than about perfect parallelism in medium or platform. People talking online about sports, movies, holidays, family, and other "normal" topics form an appropriate contrast for online talk about conspiracy theories.

Methodology

Given these data, we were ready to build ML classifier models to detect conspiracy theory language. Very broadly, ML classification is akin to teaching a computer to tell the difference between cat pictures and

---

[1] Gnip was a social media aggregator company, later purchased by Twitter and incorporated as their "Historical PowerTrack API" (undated).

dog pictures. Both animals look somewhat similar (furry, four legs, tail, two eyes, etc.); to a baby, they might be indistinguishable. But as we grow older and learn, we can tell cats from dogs (classify them) perfectly well. Human-supervised ML processes are somewhat similar and involve giving many human-labeled examples (*training data*) to an algorithm until the computer has a good model of "cat versus dog." Furthermore, because computers do not actually read or see as humans do, we have to convert all the features we use (in this case, words and stance phrases) into numbers that the computer can interpret. Additionally, it is standard in these efforts to break training data down into *train* and *test* sets: one portion of data is used to train the model, another portion is used to test the model. The hold-out test portion is labeled so that we can use performance of the model to make an estimate of how well the model might work in the wild. Although our example describes the process for classifying images, the process for classifying documents with human-supervised ML works similarly, albeit with different algorithms.

Our modeling process involved three steps. First, we established a baseline using BERT, an existing, close to state-of-the-art general language model. We then tested a stance-only model; finally, we tested a hybrid version. Here, we explain our modeling efforts in detail. (A more technical version is provided in Appendix A.)

Modeling Using Semantic Content, Stance, and a Hybrid of the Two

BERT is a relatively new and widely adopted language representation model developed at Google in 2018 (Devlin et al., 2019). It sets itself apart from previous deep-language representation models in the way that it accounts for word context, helping capture semantic relationships. Very broadly, because such words as "pet" are very often near such words as "cat," "dog," and "hamster," BERT captures the semantic relationship that all three are pets. This leads to more-robust and context-sensitive representation of semantic meaning in the model. Additionally, BERT is a general model: It has been trained primarily on Wikipedia, along with a smaller corpus of publicly available eBooks. Although Wikipedia entries are a specific genre and do not offer specialized discourse (e.g., scientific, literary, conversational), they are a broad basis for learning general word meaning. Furthermore, BERT addressed limitations of prior models that depend on recurrence, so that words (tokens) can be processed in parallel, thereby reducing training time. Ultimately, BERT is an extremely capable generalized language representation with improved learning transfer over prior language models.

Stance Model

We used the stance enrichment feature of RAND-Lex to statistically describe each document in our social media data set.[2] For this application, we used RAND-Lex to measure 119 linguistic stance variables using Term Frequency-Inverse Document Frequency (TF-IDF), which is a shorthand way of saying "what words or phrases are common in some documents but uncommon in others and thus are a good clue for classifying?" If only a few documents in a collection have such words as "rose," "lawn," "garden," and "fertilizer," that is a good hook to help an ML model recognize the class "gardening articles." Prior research has shown that stance in RAND-Lex is generally stable over time and across many genres and purposes; thus, we expected it would work out of the box for this project (Kavanagh et al., 2018). Enrichment in RAND-Lex is a kind of "vectorizing": We map the text in documents to a 119-dimension vector representing the stance-taking for each document. The resulting vector is a series of coordinates that allows us to model the various stances in a document in a 119-dimensional space.

Conspiracy Data Set

The conspiracy data set comprises 150,000 social media posts across all four conspiracy classes (alien visitation, anti-vaccination, COVID-19 origins, and WG) and one "normal" baseline class (e.g., baseball, movies, politics, everyday occurrences). Each class is sampled equally at 30,000. As part of our commitment as researchers to protecting individuals, we cannot share actual quotes that went into our train-

---

[2] RAND-Lex is RAND's proprietary text and social media analysis software platform: a scalable cloud-based analytics suite with network analytics and visualizations, a variety of text mining methods, and ML approaches. See Kavanagh et al. (2019).

ing data. The following synthetic example posts are constructed from actual data to give a better sense of what we trained our model on:

- We shouldn't be so close minded. Is the government hiding aliens' existence from us?
- Big Pharma conducted those safety trials. The evidence is in vaccine-injured kids!
- Is this a Chinese bioweapon or a Deep State virus?
- If gunmakers can be held liable for gun-related injury or death, then why can't Pharma be liable for vaccine-related injury or death?
- The same people who wanted an end to apartheid are turning a blind eye to WG. They turn a blind eye to the Zionists murdering Palestinians.

This data set is then split 80–20 between training and test sets. Training data for the stance model were cleaned with standard text-preparation techniques, such as removing emojis, most Unicode characters, punctuation, and zero padding. Although removing "stop words" (small words, such as prepositions or conjunctions) is common in text processing, we kept all words because they can provide context crucial to stance taking—for example, "and so" being tagged as a reasoning phrase.

Hybrid Conspiracy Model

Our next step was to combine the BERT and the RAND-Lex stance models, partly to improve performance and partly to provide a more interpretable model. Furthermore, we iterated our model building: first an easier model attempt to detect conspiracy theory language by topic and then a more challenging model attempt to detect the conspiratorial quality of those theories (along with a baseline of "normal" social media discourse). Ideally, we hoped the hybrid model would provide the contextual, semantic understanding of BERT while capturing rhetorical dimensions through RAND-Lex stance. For the BERT stage of our model, we used the full data set of conspiracy theory discussions—both sincere conspiracy theory promotion and rebuttals. When we moved on to the conspiratorial modeling effort, we used a smaller, curated data set with only conspiracy theory support talk.[3] Figure 3.1 gives a high-level overview of our modeling effort, wherein inputs from both BERT and stance are combined to produce trained models for both the detection of conspiracy topics (*topic prediction*) and adherence (*conspiracy prediction*).

Figure 3.1
Hybrid Model Overview

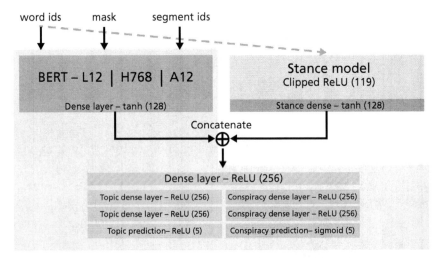

NOTES: ReLU = Rectified Linear Unit. This overview model diagram illustrates the hybrid BERT+stance model with input word identification tokenization (that is, converting human-readable text into computer-readable numbers that represent words and characters), word masking, and segment identifications (see Devlin et al., 2019). The model layers within the red region represent the stand-alone BERT model used as a baseline comparison in the following performance results. The stance model (in gray) is frozen after having been trained against the RAND-Lex output and is only capable of providing inferences. The green and yellow dense pathways (formed by multiple layers) are trained against the topic and conspiracy targets, respectively, meaning that only those layers' gradients are updated according to their targets. All layers prior to and including the hashed dense layer receive gradient updates from both the topic and conspiracy target back-propagations (neglecting the frozen stance model). This figure is explained in greater detail in Appendix A.

---

[3]   See Appendix A for illustrative examples of separating conspiracy support talk from conspiracy discussion.

Model Performance: Hybrid Model Improves Performance

In this section, we show performance for the BERT, stance, and hybrid BERT+stance models, first for detecting conspiracy theory by topic (something that existing models already do with varying success), and then by detecting the conspiratorial quality of these theories (which existing models struggle with). Results are first organized using confusion matrices,[4] which allow for a granular visualization by specific conspiracy theory; we then provide summative tables to illustrate overall model performance.

Figure 3.2 provides the confusion matrices comparing the model results for detecting conspiracy topics. Confusion matrices should be read on a diagonal from top left to bottom right, and the matrices in the figure use a heat map illustration of accuracy: Lighter greens (trending to yellow) indicate greater accuracy; darker greens (trending toward blue) indicate less accuracy. The numbers in each cell are a simple accuracy measure: the percentage accurately classified and the total number of samples tested (in parentheses). As we hypothesized, adding stance to BERT improved model performance—for example, boosting classification of alien visitation conspiracy talk from 52 percent to 87 percent.

Improving topic detection is important because social media platforms might want to measure relative volume of a given conspiracy theory—discovering that talk about a conspiracy theory is rising sharply in volume could be a valuable warning. An even more important capability would be the ability to detect and appropriately respond to content that promotes conspiracy theories. This is where our hybrid model using stance really shines, by vastly reducing the number of false positives. One important way to evaluate a model is how it reduces the number of false negatives and positives—a model that misses conspiratorial content (false negatives) but flags non-conspiratorial content (false positives) is not very useful. For this purpose, it is important to drive down false positives: Flagging useful information about COVID-19 (a kind of false positive) is the last thing social media platforms want, for example.

---

[4] A *confusion matrix* is a standard table used in ML for visualizing a model's performance. It should be read along the diagonal from top left to bottom right: Rows show the predicted class from the model; columns show the actual class.

Figure 3.2
Topic Confusion Matrices Comparison Between Models

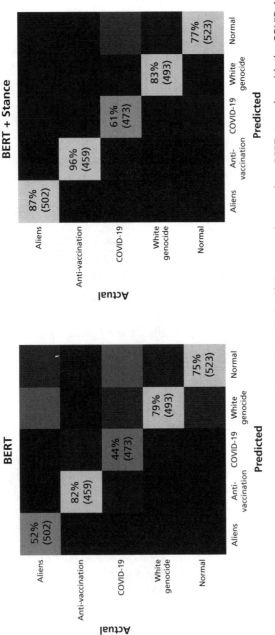

NOTES: The BERT+stance conspiracy model outperformed BERT by itself. For example, note how BERT struggles with the COVID-19 topic. This makes sense, given the breadth of subtopics that COVID-19 contains and the crossover with anti-vaccination topics. Adding stance to the model allows it to capture a signature move in COVID-19 theories: invoking people with medical authority (e.g., "Dr. [name]," "my cousin is a scientist"). These matrices, which should be read on a diagonal from top left to bottom right, are illustrated as an "accuracy" heat map: Lighter greens (trending to yellow) indicate greater accuracy; darker greens (trending toward purple) indicate less accuracy.

To test our hybrid model's performance on promotion of conspiracy theory content, we used a smaller subset of the data with only pro-conspiracy documents. Figure 3.3 illustrates both the BERT and hybrid BERT+stance confusion matrices, broken out by each conspiracy theory and a comparative "normal" social media data set. Whereas the previous confusion matrixes looked at detecting the topic of a conspiracy theory, this shows performance detecting pro-conspiracy content *within* the topics.

The following tables provide a closer look at the models' performance, using the metrics of accuracy, precision, recall, and Matthews Correlation Coefficient (MCC). In simplest terms, *accuracy* is a simple ratio of correct predictions to total predictions, *precision* is how accurate the model is when it makes a prediction, *recall* is how good the model is at not missing what it looks for, and MCC is a good overall metric for model quality. MCC is a better metric for performance when the training data does not have an equal number of samples from all classes; therefore, the second table is the most useful metric. The results of testing our stance, BERT base, and hybrid models in identifying our topics of conspiracy are shown in Table 3.1. The discrimination of whether the data are promoting conspiracy is presented in Table 3.2.

Value of Hybrid Modeling

We think that hybrid models incorporating stance offer important benefits. One benefit is the out-of-the-box improvement in performance; another is the additional interpretability to what would otherwise be a black-box model. Finally, stance greatly reduces false positives, something of particular relevance for trying to detect harmful talk online.

Stance as a Shortcut to Improving Model Performance

The BERT model comes generically pretrained and is by itself an extremely powerful tool that has been successful in modeling discourse on social media (Dadu et al., 2020; Davidson, Bhattacharya, and Weber, 2019; Mozafari, Farahbakhsh, and Crespi, 2019). However, training BERT for a specific purpose requires computing resources, time, and data, and was not part of the scope of this study. We think it is quite

Figure 3.3
Confusion Matrices by Conspiratorial Qualities

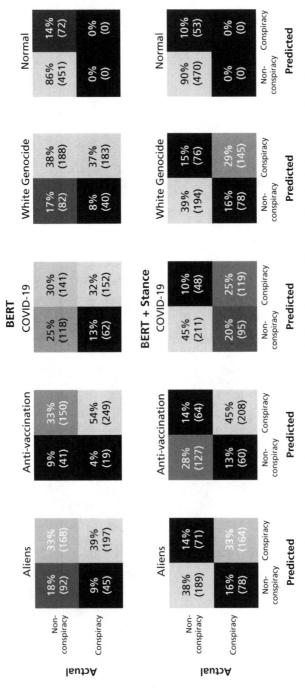

NOTES: The most noticeable improvement that arises from the inclusion of stance (illustrated in the bottom row of matrices) appears to be a reduction in false positives; however, this did result in higher false-negative predictions. The increase in false negatives hints at a subset of stance components that do not particularly describe conspiracy or non-conspiracy and are therefore common within both, thus creating a "noise floor" of incorrect predictions.

Table 3.1
Model Performance for Topics

Model	Accuracy	Precision	Recall	MCC
Stance	0.48	0.41	0.37	0.32
BERT	0.66	0.69	0.66	0.59
BERT+stance	0.79	0.79	0.79	0.74

NOTE: Performance comparison between the hybrid model and its individual model components as tested on a held-out conspiracy data set with 2,450 samples across all four conspiracy topics and a normal topic. The number of samples per topic is as follows: aliens = 502; anti-vaccination = 459; COVID-19 = 473; WG = 493; and normal = 523.

Table 3.2
Model Performance for Conspiracy

Model	Accuracy	Precision	Recall	MCC
Stance	0.42	0.41	0.98	0.06
BERT	0.64	0.52	0.82	0.35
BERT+stance	0.75	0.67	0.67	0.46

NOTE: Performance comparison between the hybrid model and its individual model components as tested on a held-out conspiracy data set with 2,450 samples across all four conspiracy topics and a normal topic. The number of samples per topic is as follows: aliens = 502; anti-vaccination = 459; COVID-19 = 473; WG = 493; and normal = 523.

possible that the BERT model could be trained eventually to achieve the performance of our hybrid model, but such training would require a significant investment in time and resources. Stance, because it draws off of a broad taxonomy of our shared repertoire for linguistic function, might act as a shortcut to hallmark features of a wide variety of text types. For example, the outreach efforts of extremist groups, such as the Islamic State in Iraq and Syria and al-Nusrah Front, appear to hinge on an urgent appeal (*insistence*) to act selflessly (*public virtue*) for the benefit of other Muslims (*social closeness*)—and there are existing stance categories that capture this argument strategy. As an example in this study, our model had a powerful detection hook in anti-vaccination conspiracies: the emphasis on public safety and children's health (*public virtues*).

Stance was a shortcut to capturing that for the model, giving a more robust conspiracy model with minimal training time and resources.[5]

Furthermore, stance is very stable over time—the same stance taxonomy that improves classification in our model also works on Elizabethan-era language (Hope and Witmore, 2010). We did not explore longitudinal drift in conspiracies, but it is our assumption that BERT might have difficulties in discriminating this because conspiracy phrases and slang change over time,[6] whereas linguistic stance would be less dependent on those same phrases or terms used. The specifics of conspiracy language—names, places, events—change relatively quickly. However, stance—talk about the past or future, certainty or uncertainty, etc.—is relatively stable; therefore, using stance might make models more durable over time.

Stance and Increased Interpretability

Stance also adds interpretability to models, helping us understand how different kinds of conspiracy discourse function. BERT and other powerful ML models are black boxes: They work well, but we cannot look under the hood to see why. Our hybrid model is explained and illustrated in more detail in the appendixes, where we demonstrate that the functional analysis of the different conspiracy theories is partly driven by feature importance output for the model.

Briefly, we believe that the more robust, interpretable model of conspiracy reflects an important advance in interpretability. Although it is quite possible that a specialized word-embedding model could be specifically trained to capture stance, our existing expert dictionaries act as a shortcut. Furthermore, the stance taxonomy of language

---

[5]   It took us approximately eight hours for each model on an NVIDIA P100 GPU platform.

[6]   BERT will have difficulty in this case because it has not learned any representations between the new and existing conspiracy terms and features. This can be adjusted with periodic fine-tuning of BERT if data containing those connections between established conspiracy terms and new terms are present. A notable example of this is within the COVID-19 data: Early conspiracy theories were centered around coronavirus escaping from the Wuhan Institute of Virology; not long afterward, additional theories sprouted that focused on Microsoft cofounder Bill Gates and installation of 5G cell towers.

from CMU represents deep domain knowledge about human linguistic behavior that allows for deep insight.

Lowering False Positives

Most notably, adding stance appears to increase the model's ability to reject false positives. This is particularly important because an over-zealous system that wrongly flags innocuous or even beneficial talk could result in more harm than good. Reducing false positives is also useful when discriminating within topics, where factual discourse on a topic includes phrases, keywords, or sentiments that overlap with conspiracy, as illustrated in our COVID-19 example. The use of stance in modeling does not need to be exclusive to conspiracy theory language; it should be capable of providing insight to other important social media topics—such as hateful speech or trolling behavior, which can also be difficult to identify.

Key Insights from Our Modeling Effort

We found that a hybrid approach to modeling conspiracy theory language worked well and offers several benefits over current approaches. One important benefit is that a hybrid BERT+stance approach appears to function as an out-of-the-box way to inductively capture genre features, obviating the need for specialized training for such generic, pre-trained models as BERT. Another is that adding stance makes it possible to create models that are more interpretable, understanding to what degree semantic content (the BERT portion) and the various stance features contribute to classification. This interpretability is critical for such tasks as dealing with conspiracy theory language, where insight is as important as performance. Finally, hybrid modeling drastically reduced false positive rates, generally cutting them in half. We think this is very important from the perspective of social media platforms that wish to avoid flagging or moderating nonharmful content. In the next chapter, we detail insights from our review of the literature on conspiracy theories online and our mixed-method analysis of the four conspiracy theories we studied.

CHAPTER FOUR

Conclusion and Recommendations

In our study, we found that a novel approach to ML could both improve performance and give us new insights into how online conspiracy theories function. By combining powerful, existing ML approaches, such as deep neural network word embeddings (e.g., BERT), with domain knowledge from linguistics and rhetorical studies (stance features), we were able to advance practice specifically in the detection of conspiracy language, with broad implications for ML classification of documents that are marked more by sociocultural meaning than semantic content.

This innovation was the direct result of Google's Jigsaw unit framing the problem as not simply a technical challenge but rather as a sociocultural challenge that required a holistic approach. We think that this sort of openness to improving ML through the creative use of insights from social science and domain experts is important in confronting the scale of difficulty around conspiracy theories specifically and around Truth Decay more broadly. We hope that other social media platforms will follow suit and embrace creative approaches to sociocultural problems that go beyond purely technical solutions.

In addition to the practical output of an improved ML model for conspiracy theories, we also synthesized the model outputs of our effort with best practices derived from existing research literature. Understanding the rhetorical function of harmful conspiracy theories can inform evidence-based interventions to reduce their adherence and spread. We close this report with recommendations for mitigating the spread and harm from online conspiracy theories.

Policy Recommendations for Mitigating the Spread of and Harm from Conspiracy Theories

Transparent and Empathetic Engagement with Conspiracists

The open nature of the social media offers numerous opportunities to engage with conspiracy theorists. These engagements should not aggravate or provoke conspiracy theory adherents. Instead of confrontation, it might be more effective to engage with conspiracists in a transparent and sensitive manner. Public health communicators recommend engagements that communicate in an open and evidence-informed way that create safe spaces to encourage dialogue, foster community partnerships, and counter misinformation with care. In particular, validating the emotional concerns of participants could encourage productive dialogue.

An additional technique beyond flagging specific conspiracy content is *facilitated dialogue*, which is when a third party facilitates communication, either in person or separated, between conflict parties (Froude and Zanchelli, 2017). This approach might help in communication between authoritative communities (such as doctors or government leaders) and conspiracy communities. Facilitated dialogues could also be carried out at lower levels in the form of facilitated discussions that help acknowledge fears and address feelings of existential threat for the participants.

Correcting Conspiracy-Related False News

One possible intervention that public health practitioners could consider is to correct instances of misinformation using such tools as real-time corrections, crowdsourced fact-checking, and algorithmic tagging. In populations that hold preexisting conspiratorial views, the evidence for the effectiveness of corrections is mixed, but results are consistently positive in studies investigating corrections of health-related misinformation in general populations.

Overall, the weight of the evidence appears in favor of such corrections. In addition, efforts to correct misperceptions in conspiracy-prone populations also should follow the advice of public health practitioners and do so in a manner that is transparent and sensitive to the concerns of pro-conspiracy audiences.

Engagement with Moderate Members of Conspiracy Groups

Conspiracists have their own experts on whom they lean to support and strengthen their views, and their reliance on these experts could limit the impact of formal outreach by public health professionals. Our review of the literature shows that one alternative might be to target outreach to moderate members of such groups who could, in turn, exert influence on the broader community. Commercial marketing programs use a similar approach when they engage social media influencers or "brand ambassadors" who then credibly communicate advantages of a commercial brand to their own audiences on social media.[1] This approach is supported by academic research suggesting that people are more influenced by their social circles than by mass communication (Guidry et al., 2015). It might be possible, for example, to convey key messages to those who are only vaccine hesitant; these individuals might, in turn, relay such messages to those in anti-vaccination social media channels.[2] Moderates who could influence WG members might be religious or political leaders or political pundits.

Addressing of Fears and Existential Threats

Underlying fears in the anti-vaccination and WG groups appear to be powerfully motivators for these groups. For anti-vaccination advocates, the fear rests on concerns about vaccine safety; for WG, that fear rests on a belief in the (perceived) existential threat to the white race. To the extent that interventions can address such fears, they might be able to limit the potential societal harms caused by both groups. Efforts that target those who are vaccine hesitant, for example, could address concerns by highlighting research on vaccine safety, the rigorous methods used in vaccine safety trials, or the alternative dangers that await those who are not vaccinated. Given that some WG conspiracists are willing to engage in rational debate and that successful persuasion requires

---

[1]   Influencer engagement programs have also been recommended as a strategy to counter violent extremism (Helmus and Bodine-Baron, 2017).

[2]   Some have not yet decided to commit to the anti-vaccine cause; others opt for some but not all vaccines; and still others prefer administering vaccines in a more gradual schedule than the Centers for Disease Control and Prevention recommends.

using the intended audience's values rather than the speaker's values (Marcellino, 2015), it might be more persuasive and effective to address claims that minorities will annihilate whites than to attempt to promote themes of racial equality.

Data and Methodology

Data

Both our modeling effort and text analysis used the same social media data set: 150,000 samples across four types of conspiracy language and a baseline sample of "normal" non-conspiracy talk. To account for differences across various social media platforms, we collected data from multiple sources: Twitter; Reddit; a variety of online forums and blogs; and some single sources, such as the transcript of the "Plandemic" (2020) video.

Our data collection was conducted through the social media tracking company Brandwatch. This gave us access to archived content from multiple social media platforms, which we culled using queries in Boolean format and by filtering results in terms of location, time, language, and other criteria. For the purposes of this study, we queried for English only.

In consultation with Jigsaw, we selected four specific conspiracy theory topics: alien visitations, anti-vaccination content, COVID-19 origins, and WG. Our social media queries were driven by an iterative approach drawing on a qualitative media assessment. Table A.1 lists the search parameters and the final sample size obtained for each of the four queries.

In addition to an iterative search strategy, we used expert input for both the anti-vaccine and COVID-19 queries. We also used the news archive service Nexis Uni to search for the terms "coronavirus AND conspiracy," limiting the dates to January 1, 2020, through April 12, 2020. Over 10,000 articles were returned, sorted by rel-

Table A.1
Data Query Parameters

Example Search Terms	Period Covered	Documents
Alien visitations aliens, ufo, extraterrestrials, visitation, roswell, tunguska, crop circles, oumuamua, annunaki, secret, government	January 1–March 31, 2020	~30,000
Anti-vaccine myths vaccine, immunization, safety, harm, thimerosal, adjuvant, big pharma, cover-up, autism, infertility	January 1–March 31, 2020	~160,000
COVID-19 coronavirus, covid, lab, secret, government, bioweapon, man-made, 5g, gates, pirbright, deep state	January 1–April 16, 2020	~60,000
White genocide white genocide, white rights	January 1–March 31, 2020	~50,000

evance (as determined by Nexis Uni). From the top 50 items, we selected 11 articles that seemed the most pertinent. These articles were manually abstracted for any COVID-19–related conspiracy theories or myths mentioned in the text.

We found that not all of our search results were useful. Some were not relevant to the topic of interest (e.g., discussions of the television show *Roswell* as opposed to actual alien visitations). Others, although relevant, did not contain content from conspiracy theory proponents: Some were neutral in tone while others mocked or tried to debunk the theories. Brandwatch (undated) has a custom ML algorithm that allows users to provide many examples of the kinds of posts they want, which allowed us to further refine our collected social media data for genuinely conspiratorial material. Perhaps because of novelty, our COVID-19 search yielded conspiracy-relevant content without this additional ML refinement step.

The data thus obtained were downloaded from Brandwatch and further scrutinized by our team for quality control. This led to additional cleansing that narrowed the results significantly. First, we discarded any items containing fewer than 2 or more than 500 words. Tests of our model showed that its performance did not significantly improve with addition of documents exceeding 300-word tokens in length; the larger items thus would add computational strain without much benefit. Second, we discarded duplicate posts—for example, those resulting from retweets or bot activity. We also filtered out social media posts consisting solely of emojis or URLs or containing non-Latin characters and tags commonly used on Reddit to indicate sarcasm ("\s" and "/s"). Last, the research team's human inspection of conspiracy speech patterns led to the addition of several filters that were developed while examining the data. For example, we discarded posts that included the use of quotation marks around the term *white genocide* and posts that used the word *conspiracy*, which was not used by actual conspiracy theory proponents.

Methodology

Literature Review Method
Approach
We first conducted a systematic review of the literature on conspiracy theories and social media as a way of placing the results of our quantitative analysis in proper context. For this review, we searched ten databases with a search string intended to gather articles addressing topics that focused on both conspiracy theories and social media.[1] We also sought papers published after 2003. The specific search string was as follows:

> (TITLE-ABS-KEY(conspira*OR"conspiracytheor*"OR"pseudo-scien*" OR {war on science} OR "anti-scien*" OR "antiscien*" OR "anti-vaxx*" OR "anti-vaccination" OR "white genocide" OR "white rights" OR "climate change denial" OR "climate change denier*" OR "global warming denial" OR "global warming

---

[1]   These databases were Pubmed, the Institute of Electrical and Electronics Engineers, Policy File, Proquest military, The Cumulative Index to Nursing and Allied Health Literature, EBSCO Military Government Collection, Psychinfo, Academic Search Complete, Web of Science, and Scopus.

denier*" OR "holocaust denier*" OR "holocaust denial" OR "replacement theory" OR "replacement theology" OR "deep state" OR qanon OR "crisis actor*" OR {crisis acting} OR "flat earth") AND TITLE-ABS-KEY ("social media" OR reddit OR twitter OR "tweet*" OR facebook OR youtube OR instagram OR whatsapp OR tiktok) AND PUBYEAR AFT 2003)

This yielded a total of 328 papers. We then reviewed titles and abstracts for this data set and identified 166 articles that appeared to meet the study's entry criteria, which was a focus on both conspiracy theories and social media, original data collection and analysis, and publication in a peer-reviewed journal. After a brief training session to promote interrater reliability, we coded the remaining 166 papers. Ordinarily, each paper would be reviewed by two independent raters who would then adjudicate differences in code application,[2] but the limited scope and resources for this study resulted instead in the lead author reviewing the applied codes to ensure consistency in code applications. Ultimately, 108 studies qualified for this review.[3]

---

[2] It is important to caveat several limitations in the methods employed in this systematic review. Best practice for systematic reviews calls for the review of titles and abstracts to be conducted by two independent raters (Okoli, 2015). Likewise, researchers should employ two separate reviewers in analyzing and coding the research papers that make up the final systematic review database. In instances where the codings of the two reviews differ, they work together alongside a third researcher to adjudicate the differences. The scope and resources for this study did not permit dual codings.

[3] Are, 2019; Arif et al., 2018; Bagavathi et al., 2019; Basch and MacLean, 2019; Basch, Milano, and Hillyer, 2019; Berkowitz and Liu, 2016; Bessi, 2016; Bessi, Caldarelli, et al., 2014; Bessi, Coletto, et al., 2015; Bessi, Petroni, et al., 2016; Bessi, Scala, et al., 2014; Bessi, Zollo, Del Vicario, Scala, Caldarelli, et al., 2015; Bessi, Zollo, Del Vicario, Scala, Petroni, et al., 2017; Bhattacharjee, Srijith, and Desarkar, 2019; Bloomfield and Tillery, 2019; Bode and Vraga, 2018; Bradshaw et al., 2019; Bradshaw et al., 2020; Brainard and Hunter, 2020; Briones et al., 2012; Broniatowski et al., 2018; Brugnoli et al., 2019; Buchanan and Beckett, 2014; Chen, Zou, and Zhao, 2019; Chin et al., 2010; Colella, 2016; Conti et al., 2017; Covolo et al., 2017; Cuesta-Cambra, Martínez-Martínez, and Niño-González, 2019; Del Vicario, Bessi, et al., 2016; Del Vicario, Vivaldo, et al., 2016; Dunn et al., 2017; Ekram et al., 2019; Essam, Aref, and Fouad, 2019; Faasse, Chatman, and Martin, 2016; Fadda, Allam, and Schulz, 2015; Farkas and Neumayer, 2020; Featherstone, 2019; Frew, 2012; Furini, 2018; Gandhi, 2020; Gesser-Edelsburg, 2018; Giese, 2020; Glenski, 2018; Golbeck et al., 2018; Greenberg, Dube, and Dreidger, 2017; Gualda and Ruas, 2019; Guidry et al., 2015; Gunaratne, Coomes, and Haghbayan, 2019; Harvey et al., 2019; He et al., 2016; Hoffman

Description of the Conspiracy Theory Literature

The vast majority of the reviewed set of publications focused on analyzing social media data with a minority of reports employing survey methods, modeling, and case study analyses (Figure 3.1). Of the studies that examined social media data, 40 percent focused on analyzing Facebook data and 30 percent focused on analyzing Twitter data. The most common focus of these studies was on anti-vaccination conspiracy theories (Figure 3.3). Possibly reflecting the newfound interest in academia of disinformation and fake news, the vast majority of studies were published in 2019 with a steady rise in publications beginning in 2012 (Figure 3.4).

Methods in the Conspiracy Theory Literature

This section had 40 studies covering topics of health (vaccines, Zika, stem cells), various conspiracy theories, environment (climate change, fluoride), and fake news. Some form of text analysis (such as linguistic, stance, topic, or sentiment) conducted either through human or machine coding was the most common research method, as follows:

- text analysis: 26
- text and network: 5
- internet search : 3

et al., 2019; Hornsey et al., 2020; Hui et al., 2018; Hussain et al., 2018; Iannelli et al., 2018; Jacques and Knox, 2016; Jamison et al., 2020; Jenkins and Moreno, 2020; Kearney et al., 2019; Keelan et al., 2010; Klein, Clutton, and Dunn, 2019; Klein, Clutton, and Polito 2018; Kou et al., 2017; Krishnendhu and George, 2019; Landrum, Olshansky, and Richards, 2019; Larson et al., 2014; Lutkenhaus, Jansz, and Bouman, 2019; Madden et al., 2012; Mahajan et al., 2019; Marcon, Murdoch, and Caulfield, 2017; Meleo-Erwin et al., 2017; Meylakhs et al., 2014; Mitra, Counts, and Pennebaker, 2016; Mocanu et al., 2015; Narayan and Preljevic, 2017; Nerghes, Kerkhof, and Hellsten 2018; Nugier, Limousi, and Lydié, 2018; Numerato et al., 2019; Okuhara et al., 2018; Okuhara et al., 2019; Penţa and Băban, 2014; Poberezhskaya, 2018; Porat et al., 2019; Porreca, Scozzari, and Di Nicola, 2020; Pyrhonen and Bauvois, 2019; Roxburgh et al., 2019; Samantray and Pin, 2019; Samory and Mitra, 2018a; Samory and Titra, 2018b; Schmitt and Li, 2019; Selepak, 2018; Sharma et al., 2017; Smith and Graham, 2019; Sommariva et al., 2018; Song and Gruzd, 2017; Stæhr, 2014; Steffens et al., 2019; Straton et al., 2019; Takaoka, 2019; Tingley and Wagner, 2017; Tomeny, Vargo, and El-Toukhy, 2017; Vijaykumar et al., 2018; Wong, Wong, and AbuBakar, 2020; Wood, 2018; Xuan and MacDonald, 2019; Yuan, Schuchard, and Crooks 2019; Zollo et al., 2015; Zollo et al., 2017.

- text and video: 3
- statistical: 2
- text analysis (other): 1.

ML Methodology
BERT

The Bidirectional Encoder Representations from Transformers is a novel language representation model developed at Google in 2018 (Devlin et al., 2019). BERT sets itself apart from previous deep language representation models in the way that its generalized bidirectional modeling uses word or token masking, leading to a more robust and context-sensitive representation. The BERT model used in this work is the uncased model with 12 transformer blocks, 12 self-attention heads, and hidden layer sizes of 768, amounting to 110 million total parameters.[4] For inclusion in a hybrid model for predicting conspiracies, two additional dense layers are added to the base BERT model. BERT by itself is an extremely capable generalized language representation, but we believe that a more robust interpretable model of conspiracy can be developed by adding in domain knowledge, such as that of linguistic stance.

Stance Proxy Model

We used the stance enrichment feature of RAND-Lex to statistically describe each document in our social media data set.[5] For this application, we used RAND-Lex to measure 119 linguistic stance variables using TF-IDF. Our hypothesis was that a generalized language representation, such as stance, can help model discourse because rhetorical function in language is stable. The RAND-Lex stance model maps input text documents to a 119-dimension vector representing the stance-taking content for each document. The components of the stance vector v_i range from $0 \leq v_i \leq 100$, with any stance vector capable of taking multiple stances simultaneously and the possibility that a

---

[4] For this study, we used the TensorFlow BERT base model (TensorFlow, undated).

[5] RAND-Lex is RAND's proprietary text and social media analysis software platform: a scalable cloud-based analytics suite with network analytics and visualizations, a variety of text-mining methods, and ML approaches. For example, see Kavanagh et al. (2019).

document takes no stances whatsoever giving a zero vector. Unfortunately, we cannot treat this as a multilabel classification; the relative magnitudes across stances are also important, and doing so would result in the loss of a significant amount of contextual information.

Building a Proxy for Stance

Our effort did not feature direct transfer of RAND-Lex's stance dictionaries to Google. Instead, a proxy model was created, moving from the TF-IDF approach to a neural network embedding model. Our proxy RAND-Lex stance model was developed in Python version 3.7 using the Keras and TensorFlow libraries and is composed of an embedding layer of input length 300, input dimension 28,869, and output dimension 128, corresponding to the maximum document length of 300 tokens and vocabulary of 28,869 unique base words. The embedding layer was flattened and then fed into three consecutive dense layers. All layers have Rectified Linear Unit (ReLU) activation functions. The final dense output layer has an output of 119, corresponding to the 119 stance components, and it uses a ReLU clipped at 100, thus satisfying the stance component value restrictions mentioned previously. Finally, because we are not dealing with probabilities and our problem is one of regression, the mean squared error is used as the loss function.

Training data for the proxy RAND-Lex model were cleaned with standard text-preparation techniques, such as removing emojis, most unicode characters, punctuation, and zero padding. We retain all stop words because they can provide context crucial to stance-taking. Capitalization of characters can provide a useful hook into describing emotion (and, potentially, intent of sarcasm), but we remove all capitalizations to reduce the complexity of this model. (Such analysis was outside the capabilities of RAND-Lex when we used it). To reduce redundant computations, the SentencePiece tokenization used by BERT is reused for input into the linguistic stance model (Kudo and Richardson, 2018).

The conspiracy data set comprises 150,000 social media documents (e.g., posts, comments, tweets) across all four conspiracy classes and one class of "normal" talk on social media, where each class is sampled equally at 30,000. This data set is then split 80–20 between training and test sets. The fidelity of the proxy RAND-Lex stance model can be seen in Figure A.1, which illustrates how the predicted stance vectors are compared with an unseen set of test stance vectors. The average and standard deviation of the error across our stance vector is 2.7 + 2.3 percent, with the stance component of 'Numbers' having the largest magnitude of error, at 8.5 percent. However, what is most important about the fidelity of this model is not the regressed magnitudes of individual stance components but the relative contributions of each stance. When looking at the relative regressed magnitudes— that is, normalizing by the largest magnitude component—the error between the predicted and test stance vectors becomes 0.03 + 0.22 percent. This is an important distinction: The stance magnitudes by themselves are somewhat arbitrary, but the linguistic stance content of one component relative to another component for the same input text is where our stance model will be most useful.

Hybrid Conspiracy Model

BERT by itself is an extremely capable language model that is suitable for many tasks, such as question answering and sentiment analysis. However, BERT does not provide much interpretability in performing these tasks, leaving the user with potentially no additional insight regarding the input text beyond the output probabilities. To provide some of that understanding, we developed a hybrid model using BERT with the proxy RAND-Lex stance model. This hybrid model is purpose-built for describing our four distinct conspiracy theories and their juxtaposition to normal social media discourse. Ideally, this hybrid model should have the contextual understanding of BERT while providing the descriptive stance-taking of RAND-Lex. An overview of the hybrid BERT+stance model can be seen in Figure A.2.

In training the hybrid conspiracy model, we had to ensure that the stance model remained unchanged because it had already been prepared through our replication of the RAND-Lex model. Therefore, for all training of the hybrid model, the underlying stance model was frozen with no gradient updates performed at any stage. For the BERT branch of our model, additional dense layers were added to perform task-specific transfer learning to our coarsest conspiracy data set,

Figure A.1
Proxy Linguistic Stance Model Error

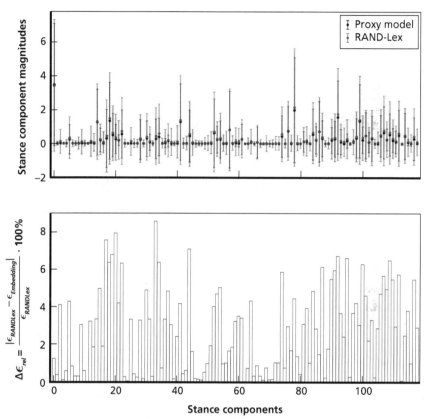

NOTES: The top image illustrates the average predicted stance component magnitudes (in red) relative to average RAND-Lex stance component magnitudes (in black). The text used for ensuring stance model fidelity is a distinct validation data set separate from the training data set. The bottom image illustrates the average relative error between the original and proxy model components where the average error is ~3 percent across all stance components.

and identifying conspiracy themes were weighted with more importance than whether the underlying data were actually describing and promoting a conspiracy. Eventually, the entire BERT branch was also frozen from gradient updates, and only dense layers downstream of

Figure A.2
Hybrid Model Overview

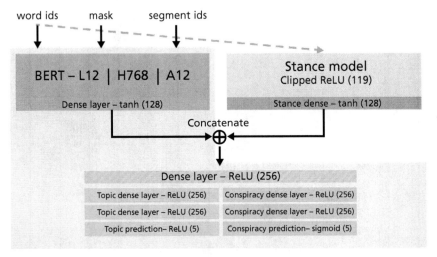

NOTES: This overview model diagram illustrates the hybrid BERT+stance model with input word identification tokenizations, word masking, and segment identifications (see Devlin et al., 2019). The model layers within the red region is the stand-alone BERT model used as a baseline comparison in the following performance results. The stance model (in gray) is frozen after having been trained against the RAND-Lex output and is only capable of providing inferences. The green and yellow dense pathways are trained against the topic and conspiracy targets, respectively, meaning that only those layers' gradients are updated according to their targets. All layers prior to and including the hashed dense layer receive gradient updates from both the topic and conspiracy target back-propagations (neglecting the frozen stance model).

the BERT+stance concatenation layer were updated through gradient descent.

For the step of moving beyond detecting talk about conspiracy theories to expressing support of conspiracy theories, we place more emphasis on a smaller, high-quality subset of the data that we hand-selected as both describing and promoting conspiracy. An example presenting the difficulty in determining promotion of conspiracy among our four topics can be seen in Figure A.3. It should be noted that the BERT stand-alone model and the hybrid model are both trained to simultaneously predict the conspiracy topic and whether the comment is promoting a conspiracy theory. This style of multitask learning has

Figure A.3
Confusion Matrices by Conspiratorial Qualities

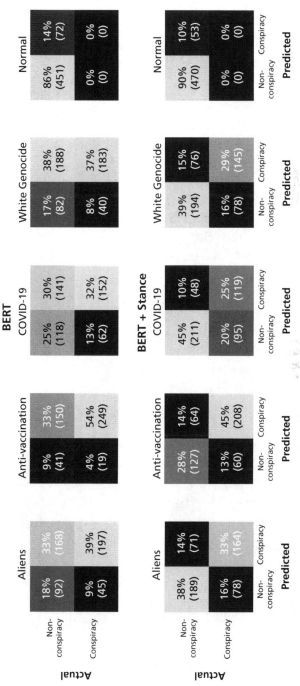

NOTES: The most noticeable improvement from the inclusion of stance (illustrated in the bottom row of matrices) appears to be a reduction in false positives; however, this resulted in higher false-negative predictions. The increase in false negatives hints at a subset of stance components that do not particularly describe conspiracy or non-conspiracy and are therefore common within both, thus creating a "noise floor" of incorrect predictions.

shown to be beneficial through a model leveraging hints provided by the simpler task (topic) to improve performance of the more difficult task (conspiracy promotion) (Ruder, 2017).

Insights

The pretrained BERT model by itself is an extremely powerful tool that has been successful in modeling discourse in social media (Dadu et al., 2020; Davidson, Bhattacharya, and Weber, 2019; Mozafari, Farahbakhsh, and Crespi, 2019). However, BERT lacks interpretability regarding why a social media comment might appear in both normal discourse and conspiracy discourse. We have shown that adding linguistic stance into a generalized language representation, such as BERT, can allow the leverage of some domain knowledge to develop a more robust model of conspiracy discourse in social media. More important, we have demonstrated that, although BERT is most likely capable of representing conspiracy language given significantly more computational or data resources, the inclusion of the stance model appears to bypass those requirements. Most notably, adding stance appears to increase the model's ability to reject false positives. This is particularly useful when discriminating within topics, where factual discourse on a topic uses phrases, keywords, or sentiment that might overlap with conspiracy discourse, as occurred in our COVID-19 example. The use of stance in this way should not be exclusive to conspiracy research; it should be capable of providing insight to other important social media topics, such as hateful discourse or trolling behavior that can also be difficult to identify.

Next Steps

Future methods for improving the BERT and linguistic stance hybrid model should focus on larger, higher-quality training data sets across an even wider sampling of social media platforms (such as Facebook, YouTube, Instagram, 4chan, and others). One particular data quality issue is the contamination of conspiracy discourse through sarcasm or quotation. Determining whether certain social media comments are sarcastic can be particularly confusing even for humans, especially without context of the greater conversation. Removal of sarcastic discourse could reduce the signal-to-noise ratio between conspiracy and non-conspiracy discussions, providing a much clearer view of the characteristic stance found in conspiracies propagated on social media.

Additionally, looking more broadly at social media platforms could provide a more robust stance profile of the conspiracies while providing a test for platform invariance. This could account for conspiracy discourse that presents itself differently across platforms (i.e., platforms that welcome conspiracy openly and those that require use of code or subtle probing for acceptance of an idea). Our research did not explore longitudinal drift in conspiracies, but it is our assumption that BERT might have difficulties in discriminating because conspiracy phrases or slang changes over time—whereas linguistic stance would be less dependent on the same phrases or terms being used. Additionally, many social media platforms rely on multimedia; for example, Twitter comments are quite often accompanied by images or videos. However, using all of this information for multimodal prediction is nontrivial. Therefore, identifying conspiracy promotion on certain platforms that are more dependent on multimedia, such as Twitter, could benefit from a multimodal model for conspiracy discrimination.

Finally, we think future work should address adversarial artificial intelligence. If social media platforms use improved text classification to lessen the harm of conspiracy theory content, it is quite possible that malign actors will seek to use algorithmic means to continue to spread them. In this sense, there is a possible arms race of defensive and offensive technology; therefore, we think that future work in this space must account for the possibility of machine means to scalably disguise or generate conspiracy theory content that fools defensive means but still communicates effectively to human readers.

Stance: Text Analysis and Machine Learning

Text Analysis Using Stance

In Chapter Two, we shared the results of a stance comparison analysis of the conspiracy theory data sets. For this task, we used RAND-Lex, RAND's proprietary, in-house text analysis suite. Stance relies on contrasting word and phrase counts of several different language categories across texts to draw meaningful insights. These words and phrases are organized into 15 linguistic parent categories (including emotions, public values, academic language, and reporting). These are further subdivided into 119 linguistic characteristics (including apology, social responsibility, citing sources, and causality). For a complete list and examples of each linguistic category and characteristic, see Table B.1.

Table B.1
Definitions of Stance Variables and Categories

Name	Definition
Personal perspective	Language from a subjective perspective, including our personal certainty, intensity, and temporal experience
First person	Self-reference (e.g., first person—I, me, my, myself)
Personal disclosure	Self-reference (e.g., first person—I, me, my) combined with personal thought or feeling verbs (e.g., I think, I feel, I believe)
Personal reluctance	First-person resistance in decisionmaking, (e.g., I am sorry that, I'm afraid that)
Autobiography	Self-reference (e.g., first person—I, me, my) combined with "have" or "used to," signaling personal past (e.g., I have always, I used to)
Personal thinking	Words indicating the unshared contents of an individual mind (e.g., believe, feel, conjecture, speculate, pray for, hallucinate); a front row seat into someone else's mind
General disclosure	Disclosing private information (e.g., confess, acknowledge that, admit, let on that, let it slip that)
Certainty	E.g., "for sure," "definitely"
Uncertainty	E.g., "maybe," "perhaps"
Intensity	Involved and committed to the ideas being expressed (e.g., very, fabulously, really, torrid, amazingly)
Immediacy	E.g., "right now," "now," "just then"
Subjective talk	Acknowledging that a perception is subjective/tentative (e.g., it seems, appears to be)
Time	Temporality, including temporal perspective
Subjective time	Experiencing time from the inside (e.g., seems like only yesterday)
Looking ahead	Words indicating the future (e.g., in order to, look forward to, will be in New York)
Predicting the future	Confident predictions, often using epistemic modals (e.g., we will, there will be)
Looking back	Mental leap to the past (e.g., used to, have been, had always wanted)

Table B.1—Continued

Name	Definition
Past looks ahead	Presents what the future looked like from the vantage of the past (e.g., Lincoln was to look for the general who could win the war for him)
Time shift	E.g., "next week," "next month"
Time duration	Temporal intervals (e.g., "for two years" "over the last month")
Biographical time	Life milestones (e.g., "in her youth" "it would be the last time")
Time date	E.g., "June 5, 2000"
Emotion	Affective language
General positivity	Covers all positive emotion language (e.g., "joy," "wonderful")
General negativity	Negative language that doesn't fall into Anger, Fear, Sadness, Reluctance, or Apology (e.g., "that sucks," "suicidal")
Anger	Words referencing anger
Fear	Words referencing fear
Sadness	Sadness (negative emotion)
Reluctance	Resistance within the mind (e.g., regret that, sorry that, afraid that)
Apology	E.g., "I'm sorry," "I have failed"
Descriptive language	Descriptions of the world
Dialogue	Dialogue cues (e.g., quote marks, "she said")
Oral talk	Oral register (e.g., "well," "uh," "um")
Concrete properties	Words indicating concrete properties (e.g., pink, velvety) revealed by the five senses
Concrete objects	Concrete nouns (e.g., table, chair)
Spatial relations	E.g., "nearby," "away from"
Scene shift	Shifts in spatial location (e.g., "left the room," "went outdoors")

Table B.1—Continued

Name	Definition
Motion	E.g., run, skip, jump
Interpersonal relationships	Text about/constructing the social world
Promises	Words indicating a promise being made (e.g., promise, promised that)
First-person promises	Words indicating a promise being made by the speaker (e.g., "I promise," "we promised that")
Reassure	Reassuring words (e.g., don't worry, it's okay).
Reinforce	Positive social reinforcement (e.g., "congratulations," "good going")
Acknowledging	Words that give public notice of gratitude to persons (e.g., I acknowledge your help, thank you); acknowledgments without gratitude are in Apology and Concessions
Agreement	Public notice of acceptance or agreement (e.g., I accept, I agree)
Social closeness	Language of social belonging, fellow feeling, or like-mindedness
Positive attribution	Positive attributions to people (e.g., "given credit for," or "ability to")
Social distancing	Negative/distant social relations (e.g., condemn, denounce, criticize)
Negative attribution	Negative attributions to people (e.g., "be unqualified for," "oafish," "psycho")
Confront	To confront or threaten the addressee (e.g., "Let's face it" "how dare you")
Public values	Language about public virtue
Public virtue	The positive, publicly endorsed values and standards of the culture (e.g., justice, happiness, fairness, "human rights")
Innovations	Significant discovery (e.g., breakthrough, cutting-edge, state-of-the-art)
Public vice	Standards and behavior publicly rejected by the culture (e.g., injustice, unhappiness, unfairness, "civil rights violations")
Social responsibility	The language of public accountability (e.g., to "take care of" our vets, or to "take on" Wall Street)

Table B.1—Continued

Name	Definition
Public language	Public sharing of talk
Rumors and media	Words circulating over formal or informal media channels. Includes institutional networks but also rumor, gossip, buzz, and memes (distinct from "authority sources")
Authority sources	Public or institutional authorities, already familiar and respected in the culture (e.g., "founding fathers," "the courts," "the Prophet," "duly authorized")
Popular opinions	Beliefs, ideas, and approaches circulating in the culture and well known. (e.g., "some hold that," "others believe that," "in the history of")
Confirming opinions	Agreeing with and supporting ideas that are already out in the culture and well known (e.g., "I recognize that," "I agree with")
Academic language	Academic register
Abstract concepts	E.g., "gross amount, "money," "evolutionary theory"; includes Latinate/Hellenic (tion, sion, ment, ogy, logy) suffixes and other patterns indicating abstract general concepts
Communicator role	Formal communication situation (e.g., "speaker" "listener" "audience")
Linguistic references	References to language (e.g., noun, verb, adjective, play, novel, poem, prose)
Citing precedent	Referencing a chain of historical decisions to which you can link your own ideas (e.g., has long been; has a long history)
Citing sources	External sources (e.g., "according to," "sources say," "point out that")
Undermining sources	Citation that hints at a biased or deficient source
Countering sources	Citation used to counter a previous statement
Speculative sources	Citation of a source that is guessing
Authoritative source	Citation of a source that knows (e.g., "support that")
Contested source	Citation of a source in a debate ("argue for," "content that")
Attacking sources	Citation attacking a previous source
Quotation	Use of quotations

Table B.1—Continued

Name	Definition
Metadiscourse	Navigational guides through the stream of language (e.g., to clarify, just to be brief, this paper will argue, my purpose is)
Reasoning	Logic and argument language
Reason forward	Chain of thought moving forward from premise to conclusion, cause to effect (e.g., thus, therefore)
Reason backward	Words indicating a chain of thought moving backward from conclusion to premise, effect to cause (e.g., because, owing to the fact, on the grounds that)
Direct reasoning	Words that initiate and direct another's reasoning (e.g., suppose that, imagine that)
Supporting reasoning	Words indicating support or evidence for a reasoning process that you or someone you are citing has
Contingency	Words indicating contingency (e.g., if, possibly)
Denial	"Not" or other negative elements in front of an assertion, denying what a listener or reader might believe (e.g., "not what you think")
Concessions	Acknowledging weaknesses in one's own position or the strengths in the position of an opponent, (e.g., although, even if, it must be acknowledged)
Resistance	Opposition or struggle between competing ideas, events, forces, or groups (e.g., "veto," "counterargument," "military operations against")
Interactions	Linguistic interactions
Curiosity	Involving the audience in a common line of thinking (e.g., what then shall we make of?)
Question	The use of questions
Future question	The use of questions that start with the epistemic modal "will" to indicate the question pertains to a future state
Formal query	The use of survey-type queries ("if so, when," or "do you know")
Attention grab	Summoning another's attention (e.g., let us, I advise you, I urge)
Your attention	Summoning the attention of a second person "you" (e.g., "look, you")

Table B.1—Continued

Name	Definition
You reference	The use of words referencing a second person "you" (you see it's good).
Request	The use of words that make requests (e.g., I request)
Follow up	Referencing a previous interaction (e.g., in response to your, per your last message)
Feedback	The use of words indicating generic feedback to another (e.g., "okay")
Positive feedback	Words indicating positive feedback (e.g., that's very good, very nice)
Negative feedback	The use of words indicating negative feedback (e.g., that's awful, that's crummy)
Prior knowledge	Indicating that ideas under discussion are already public and familiar (e.g., "as you know")
Elaboration	Adding details or explication to talk
Generalization	Indicating generalizations to members of a class (e.g., all, every)
Example	Indicating an example (e.g., "for example")
Exceptions	Exception to general states (e.g., "an exception," "the only one to")
Comparison	Indicating conceptual similarity and difference, like "more" or "fewer"
Resemblances	The use of words indicating perceptual similarity, (e.g., resembles, looks like).
Specifiers	Indicating more-specific or restricted information is to come (e.g., in particular, more specifically).
Definition	Indicating definitions (e.g., is defined as, the meaning of the term)
Numbers	The use of words indicating numbers
Reporting	Reporting states, events, and changes, as well as reporting causal sequences
Reporting states	Using verbs "is" "are" and "be" to report constant states of information, along with other reporting verbs (e.g., is carried by, is housed in)

Table B.1—Continued

Name	Definition
Reporting events	Reporting event information, usually with verbs (e.g., established, instituted, influenced).
Recurring events	The use of words reporting event recurrence over time (e.g., again, recurred)
Generic events	Reporting events that repeat over time through processes of biology, culture, and convention (e.g., sleeping, playing, working, relaxing)
Sequence	Sequential processes unfolding over time (e.g., first, second)
Mature process	Mature sequential processes, now in its late or advanced stages of development (e.g., "so thoroughly," "well into")
Causality	The causes of sequences unfolding over time (e.g., "the aftereffects," "because of," "by means of")
Consequence	The consequences in sequences unfolding over time (e.g., "resulting in," "significant effects")
Transformation	E.g., "broke off, "came true," "metamorphize"
Substitution	E.g., "in exchange for," "in place of"
Updates	Reporting an update (e.g., "have now," "announced that")
Precedent setting	The use of words reporting historical "firsts"
Directions	Directions and guidance
Imperatives	Imperative verbs (usually beginning a sentence (e.g., "need to respond by" ". Stop" ". Take your")
Procedures	Procedures to perform (e.g., "go back to step" "use only")
Body movement	Physical directions to the body (e.g., clasp, grab, twist, lift up)
Confirmed experience	Confirmation of a just-taken instructional step (e.g., "as you see" "now you will")
Error recovery	How to recover from error (e.g., "should you get lost" "if that doesn't work")
Insistence	Insistence, either on action (e.g., you need to come) or on reasoning (e.g., you need to consider). Hallmarks of insistence language are the modals "must" "should" "need" and "ought"
Prohibition	E.g., "ought not" "should never"

Table B.1—Continued

Name	Definition
Narrative	Story-telling
Narrative verbs	The use of past -ed verbs indicating the action of a story (e.g., came, saw, conquered)
Asides	Side comments or return from one (e.g., "by the way" "anyway")
Characters	Characters in the social and physical world
Person pronoun	Pronouns (e.g., he, she) indicating persistent topical reference to people, especially in narrative
Neutral attribution	She/he/they attributions without positive or negative assessment (e.g., "he works" "she testified" "they vanished")
Personal roles	Referencing a person's formal occupational and identity roles (e.g., "butcher" "African American" "veteran")
Dialogue cues	E.g., "he added" "piped up"
Oral cues	E.g., "you guys" "ROFL"

NOTE: The 119 stance categories in RAND-Lex are derived from a rhetorical taxonomy of language developed by CMU, representing an expert dictionary with many millions of entries—the example words or phrases and descriptions provided here are meant to help illustrate each of the 119 stances measured.

RAND-Lex's stance analysis module allows human text analysts to compare text corpora through statistical tests (i.e., Tukey's Honest Statistical Difference). In addition, analysts can run principal component analysis to infer latent structure in texts from statistically significant covariance of language features.[1] *Stance comparison* combines machine-reading of text with human interpretation of feature-rich examples of patterns in text data. For example, our stance comparison showed that use of *concrete properties*—sense words, such as green,

---

[1] RAND-Lex's stance comparison uses two main statistical tests: Analysis of Variance (ANOVA) Tukey's Honest Statistical Difference and principal component analysis. Tukey's test compares significantly different means from ANOVA for pairwise comparisons between one corpus and another. Principal component analysis is a kind of data reduction that shows just what variables help distinguish data sets, including covariance of language features. This approach can expose latencies, such as rhetorical or argument strategies.

soft, or blazing hot—was a defining feature of WG talk. RAND-Lex allows analysts to quickly dive into many examples of text with a given feature for human understanding, in this case (perceived) racial categories: "*white* women" and "*brown* people."

We conducted stance comparisons for the four tested communities using the same data set used in our modeling effort: alien visitation anti-vaccination content, COVID-19 origins, and WG. We compared these groups with each other and with a baseline collection of "normal" Twitter talk. The conspiracy collections contain not only Twitter posts but also blogs and other online posts, so the baseline corpus that is drawn only from Twitter is not a perfect comparison. However, it is a useful comparison: Twitter was a major contributor to each conspiracy collection.

Using Feature Importance to Better Understand How Our Model Detects Conspiracy Theories

The previous section laid out stance variables and stance comparison in detail. These same stance variables used in our text analysis were also used as features in our model building. *Stance* is an interpretable taxonomy of language moves with a human-manageable number of variables; thus, it supports human interpretation of ML models. We leverage this by outputting the relative importance of each stance feature. For example, anti-vaccination talk is most strongly detected in our model (relative to normal discourse) using *public vices*—the bad things we want to avoid in the public sphere, such as liability, injury, or death. Likewise, WG conspiracy talk was most strongly signaled in our model by first-person constructions—for example, me, I, and my. These are not only powerful features for ML classification, they help us understand what is distinct and thus potentially meaningful about a given conspiracy theory's rhetorical moves.

We wish to emphasize that our analysis in Chapter Two of the rhetorical characteristics of different conspiracy theories is distinct from the description here of the rhetorical features that our model uses to detect conspiracy theory language from normal discourse. The former is based on a statistical analysis of significant differences among the various conspiracy theories (i.e., how conspiracy theory X functions in contrast to other conspiracy theories). Here, we are asking what features an ML model can use to detect conspiracy theory talk and support.

Figures B.1 through B.4 show the top ten most-important stance features in the model for each type of conspiracy talk when compared with normal social media discourse. The error bars in each figure show the standard deviation for two populations (e.g., aliens and normal) which is then used to calculate the standard error of the relative ratio (i.e., $X\_Aliens/X\_normal - 1$). This is calculated:

$$\sigma_f \approx |f| \sqrt{\left(\frac{\sigma_A}{A}\right)^2 + \left(\frac{\sigma_B}{B}\right)^2 - 2\frac{\sigma_{AB}}{AB}}$$

Where $f = A/B - 1$
$\quad A = $ Aliens
$\quad B = $ Normal.

Figure B.1
Aliens Language Category Stance Importance

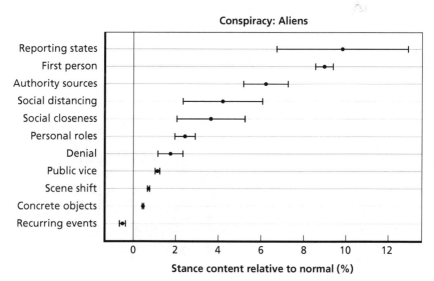

NOTE: This figure presents a within-corpus comparison of the distribution of stance categories for alien visitation discourse relative to a normal discourse data set, showing the most statistically significant stance components that differ from normal discourse seen in social media.

Figure B.2
Anti-Vaccination Language Category Stance Importance

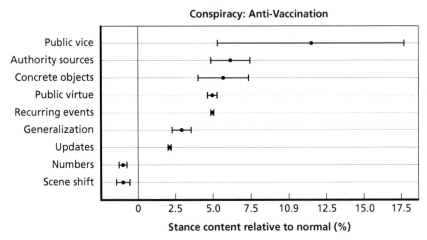

NOTE: This figure presents a within-corpus comparison of the distribution of stance categories for anti-vaccination discourse relative to a normal discourse data set, showing the most statistically significant stance components that differ from normal discourse seen in social media.

Figure B.3
COVID-19 Language Category Stance Importance

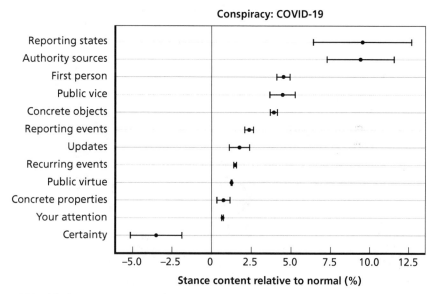

NOTE: This figure presents a within-corpus comparison of the distribution of stance categories for COVID-19 discourse relative to a normal discourse data set, showing the most statistically significant stance components that differ from normal discourse seen in social media.

Figure B.4
White Genocide Language Category Stance Importance

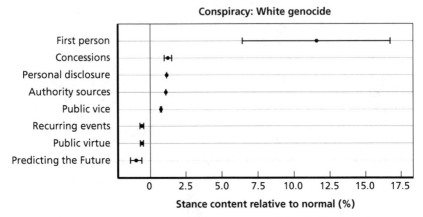

NOTE: This figure presents a within-corpus comparison of the distribution of stance categories for WG discourse relative to a normal discourse data set, showing the most statistically significant stance components that differ from normal discourse seen in social media.

References

André, F. E., "Vaccinology: Past Achievements, Present Roadblocks and Future Promises," *Vaccine*, Vol. 21, No. 7–8, 2003, pp. 593–595.

Are, C., "Patterns of Media Coverage Repeated in Online Abuse on High-Profile Criminal Cases," *Journalism*, October 14, 2019.

Arif, N., M. Al-Jefri, I. H. Bizzi, G. B. Perano, M. Goldman, I. Haq, K. L. Chua, M. Mengozzi, Neunez, H. Smith, H., and P. Ghezzi, "Fake News or Weak Science? Visibility and Characterization of Anti-Vaccine Webpages Returned by Google in Different Languages and Countries," *Frontiers in Immunology*, Vol. 9, June 2018, Article 1215.

Bagavathi, A., P. Bashiri, S. Reid, M. Phillips, and S. Krishnan, "Examining Untempered Social Media: Analyzing Cascades of Polarized Conversations," *Proceedings of the 2019 IEEE/ACM International Conference on Advances in Social Networks Analysis and Mining*, Vancouver, Canada, 2019, pp. 625–632.

Banas, J. A., and G. Miller, "Inducing Resistance to Conspiracy Theory Propaganda: Testing Inoculation and Metainoculation Strategies," *Human Communication Research*, Vol. 39, 2013, pp. 184–207.

Basch, C. H., and S. A. MacLean, "A Content Analysis of HPV Related Posts on Instagram," *Human Vaccines and Immunotherapeutics*, Vol. 15, No. 7–8, 2019, pp. 1476–1478.

Basch, C. H., N. Milano, and G. C. Hillyer, "An Assessment of Fluoride Related Posts on Instagram," *Health Promotion Perspectives*, Vol. 9, No. 1, 2019, pp. 85–88.

Berkowitz, D., and Z. M. Liu, "Media Errors and the 'Nutty Professor': Riding the Journalistic Boundaries of the Sandy Hook Shootings," *Journalism*, Vol. 17, No. 2, 2016, pp. 155–172.

Bessi, A., "Personality Traits and Echo Chambers on Facebook," *Computers in Human Behavior*, No. 65, 2016, pp. 319–324.

Bessi, A., G. Caldarelli, M. Del Vicario, A. Scala, and W. Quattrociocchi, "Social Determinants of Content Selection in the Age of (Mis)Information," in L. M. Aiello and D. McFarland, eds, *Social Informatics*, Cham, Switzerland: Springer Lecture Notes in Computer Science, Vol. 8851, 2014, pp. 259–268.

Bessi, A., M. Coletto, G. A. Davidescu, A. Scala, G. Caldarelli, and W. Quattrociocchi, "Science vs Conspiracy: Collective Narratives in the Age of Misinformation," *PLoS ONE*, Vol. 10, No. 2, 2015, Article e0118093.

Bessi, A., F. Petroni, M. D. Vicario, F. Zollo, A. Anagnostopoulos, A. Scala, G. Caldarelli, and W. Quattrociocchi, "Homophily and Polarization in the Age of Misinformation," *European Physical Journal: Special Topics*, Vol. 225, No. 10, 2016, pp. 2047–2059.

Bessi, A., A. Scala, L. Rossi, Q. Zhang, and W. Quattrociocchi, "The Economy of Attention in the Age of (Mis)Information," *Journal of Trust Management*, Vol. 1, No. 1, 2014, pp. 1–13.

Bessi, A., F. Zollo, M. Del Vicario, A. Scala, G. Caldarelli, and W. Quattrociocchi, "Trend of Narratives in the Age of Misinformation," *PLoS ONE*, Vol. 10, No. 8, 2015, Article e0134641.

Bessi, A., F. Zollo, M. Del Vicario, A. Scala, F. Petroni, B. Gonçalves, and W. Quattrociocchi, "Everyday the Same Picture: Popularity and Content Diversity," in B. Gonçalves, R. Menezes, R. Sinatra, and V. Zlatic, eds., *Complex Networks VIII*, Cham, Switzerland: Springer, *Springer Proceedings in Complexity*, 2017, pp. 225–236.

Bhattacharjee, U., P. K. Srijith, and M. S. Desarkar, "Leveraging Social Media Towards Understanding Anti-Vaccination Campaigns," 11th International Conference on Communication Systems and Networks, COMSNETS, Bangalore, India, 2019.

Bloomfield, E. F., and D. Tillery, "The Circulation of Climate Change Denial Online: Rhetorical and Networking Strategies on Facebook," *Environmental Communication*, Vol. 13, No. 1, 2019, pp. 23–34.

Bode, L., and E. K. Vraga, "See Something, Say Something: Correction of Global Health Misinformation on Social Media," *Health Communication*, Vol. 33, No. 9, 2018, pp. 1131–1140.

Braddock, K. "Vaccinating Against Hate: Using Attitudinal Inoculation to Confer Resistance to Persuasion by Extremist Propaganda," *Terrorism and Political Violence*, November, 2019.

Bradshaw, A. S., S. S. Shelton, E. Wollney, D. Treise, and K. Auguste, "Pro-Vaxxers Get Out: Anti-Vaccination Advocates Influence Undecided First-Time, Pregnant, and New Mothers on Facebook," *Health Communication*, January 9, 2020.

Bradshaw, S., P. N. Howard, B. Kollanyi, and L.-M. Neudert, "Sourcing and Automation of Political News and Information over Social Media in the United States, 2016–2018," *Political Communication*, Vol. 37, No. 2, 2019, pp. 173–193.

Brainard, J., and P.R. Hunter, "Misinformation Making a Disease Outbreak Worse: Outcomes Compared for Influenza, Monkeypox, and Norovirus," *Simulation-Transactions of the Society for Modeling and Simulation International*, Vol. 10, 2020.

Brandwatch, homepage, undated. As of March 12, 2021: https://www.brandwatch.com/

Briones, R., X. Nan, K. Madden, and L. Waks, "When Vaccines Go Viral: An Analysis of HPV Vaccine Coverage on YouTube," *Health Communication*, Vol. 27, No. 5, 2012, pp. 478–485.

Broniatowski, D. A., A. M. Jamison, S. Qi, L. AlKulaib, T. Chen, A. Benton, S. C. Quinn, and M. Dredze, "Weaponized Health Communication: Twitter Bots and Russian Trolls Amplify the Vaccine Debate," *American Journal of Public Health*, Vol. 108, No. 10, 2018, pp. 1378–1384.

Brugnoli, E., M. Cinelli, W. Quattrociocchi, and A. Scala, "Recursive Patterns in Online Echo Chambers," *Scientific Reports*, Vol. 9, No. 1, 2019, Article 20118.

Buchanan, R., and R. D. Beckett, "Assessment of Vaccination-Related Information for Consumers Available on Facebook," *Health Information and Libraries Journal*, Vol. 31, No 3, 2014, pp. 227–234.

Byrne, S., and P. S. Hart, "The Boomerang Effect: A Synthesis of Findings and a Preliminary Theoretical Framework," *Annals of the International Communication Association*, Vol. 33, No. 1, 2009, pp. 3–37.

Chen, X., L. Zou, and B. Zhao, "Detecting Climate Change Deniers on Twitter Using a Deep Neural Network," *Proceedings of the 2019 11th International Conference on Machine Learning and Computing*, Zhuhai, China, February 2019, pp. 204–210.

Chin, A., J. Keelan, G. Tomlinson, V. Pavri-Garcia, K. Wilson, and M. Chignell, "Automated Delineation of Subgroups in Web Video: A Medical Activism Case Study," *Journal of Computer-Mediated Communication*, Vol. 15, No. 3, 2010, pp. 447–464.

Clayton, K., S. Blai, J. A. Busam, S. Forstner, J. Glance, G. Green, A. Kawata, A. Kovvuri, J. Martin, E. Morgan, M. Sandhu, R. Sang, R. Scholz-Bright, A. T. Welch, A. G. Wolff, A. Zhou, and B. Nyhan, "Real Solutions for Fake News? Measuring the Effectiveness of General Warnings and Fact-Check Tags in Reducing Belief in False Stories on Social Media," *Political Behavior*, February 2019.

Colella, C., *Distrusting Science on communication Platforms: Socio-Anthropological Aspects of the Science-Society Dialectic Within a Phytosanitary Emergency*, International Conference on Language Resources and Evaluation, 2016.

Conti, M., D. Lain, R. Lazzeretti, G. Lovisotto, and W. Quattrociocchi, *It's Always April Fools' Day!: On the Difficulty of Social Network Misinformation Classification via Propagation Features*, ArXiv.org, 2017. As of March 15, 2021: https://arxiv.org/abs/1701.04221

Cook, J., S. Lewandowsky, and U. K. Ecker, "Neutralizing Misinformation Through Inoculation: Exposing Misleading Argumentation Techniques Reduces Their Influence," *PLoS One*, Vol. 12, 2017.

Covolo, L., E. Ceretti, C. Passeri, M. Boletti, and U. Gelatti, What Arguments on Vaccinations Run Through YouTube Videos in Italy? A Content Analysis," *Human Vaccines and Immunotherapeutics*, Vol. 13, No. 7, 2017, pp. 1693–1699.

Cuesta-Cambra, U., L. Martínez-Martínez, and J. I. Niño-González, "An Analysis of Pro-Vaccine and Anti-Vaccine Information on Social Networks and the Internet: Visual and Emotional Patterns," *Profesional de la Informacion*, Vol. 28, No. 2, 2019, Article e280217.

Dadu, Tanvi, Tanvi Dadu, Kartikey Pant, and Radhika Mamidi, *BERT-Based Ensembles for Modeling Disclosure and Support in Conversational Social Media Text*, ArXiv.org, June 2020. As of March 15, 2021: http://arxiv.org/abs/2006.01222

Davidson, Thomas, Debasmita Bhattacharya, and Ingmar Weber, *Racial Bias in Hate Speech and Abusive Language Detection Datasets*, ArXiv.org, May 2019. As of March 15, 2021: http://arxiv.org/abs/1905.12516

Del Vicario, M., A. Bessi, F. Zollo, F. Petroni, A. Scala, G. Caldarelli, H. E. Stanley, and W. Quattrociocchi, "The Spreading of Misinformation Online," *Proceedings of the National Academy of Sciences of the United States of America*, Vol. 113, No. 3, 2016, pp. 554–559.

Del Vicario, M., G. Vivaldo, A. Bessi, F. Zollo, A. Scala, G. Caldarelli, and W. Quattrociocchi, "Echo Chambers: Emotional Contagion and Group Polarization on Facebook," *Scientific Reports*, Vol. 6, 2016, Article 37825.

Devlin, Jacob, Ming-Wei Chang, Kenton Lee, and Kristina Toutanova, *BERT: Pre-Training of Deep Bidirectional Transformers for Language Understanding*, ArXiv.org, 2019. As of March 15, 2021: https://arxiv.org/abs/1810.04805

Dunn, A. G., D. Surian, J. Leask, A. Dey, K. D. Mandl, and E. Coiera, "Mapping Information Exposure on Social Media to Explain Differences in HPV Vaccine Coverage in the United States," *Vaccine*, Vol. 35, No. 23, 2017, pp. 3033–3040.

Ekram, S., K. E. Debiec, M. A. Pumper, and M. A. Moreno, "Content and Commentary: HPV Vaccine and YouTube," *Journal of Pediatric and Adolescent Gynecology*, Vol. 32, No. 2, 2019, pp. 153–157.

Essam, B. A., M. M. Aref, and F. Fouad, "When Folkloric Geopolitical Concerns Prompt a Conspiratorial Ideation: The Case of Egyptian Tweeters," *GeoJournal*, Vol. 84, No. 1, 2019, pp. 121–133.

Faasse, K., C. J. Chatman, and L. R. Martin, "A Comparison of Language Use in Pro- and Anti-Vaccination Comments in Response to a High Profile Facebook Post," *Vaccine*, Vol. 34, No. 47, 2016, pp. 5808–5814.

Fadda, M., A. Allam, and P. J. Schulz, "Arguments and Sources on Italian Online Forums on Childhood Vaccinations: Results of a Content Analysis," *Vaccine*, Vol. 33, No. 51, 2015, pp. 7152–7159.

Farkas, J., and C. Neumayer, "Mimicking News: How the Credibility of an Established Tabloid Is Used When Disseminating Racism," *Nordicom Review*, Vol. 41, No. 1, 2020, pp. 1–17.

Featherstone, J. D., R. A. Bell, and J. B. Ruiz, "Relationship of People's Sources of Health Information and Political Ideology with Acceptance of Conspiratorial Beliefs About Vaccines," *Vaccine*, Vol. 37, No. 23, 2019, pp. 2993–2997.

Frew, P. M., J. E. Painter, B. Hixson, C. Kulb, K. Moore, C. del Rio, A. Esteves-Jaramillo, and S. B. Omer," Factors Mediating Seasonal and Influenza A (H1N1) Vaccine Acceptance Among Ethnically Diverse Populations in the Urban South," *Vaccine*, Vol. 30, No. 28, 2012, pp. 4200–4208.

Froude, Jack, and Michael Zanchelli, What Works in Facilitated Dialogue Projects, Washington, D.C.; United States Institute of Peace, 2017.

Furini, M., and G. Menegoni, "Public Health and Social Media: Language Analysis of Vaccine Conversations," International Workshop on Social Sensing, Orlando Fla., 2018.

Gallagher J., and H. Y. Lawrence, "Rhetorical Appeals and Tactics in *New York Times* Comments About Vaccines: Qualitative Analysis," *Journal of Medical Internet Research*, Vol. 22, No. 12, 2020. As of December 20, 2020: https://www.jmir.org/2020/12/e19504

Gandhi, C. K., J. Patel, and X. Zhan, "Trend of Influenza Vaccine Facebook Posts in Last 4 Years: A Content Analysis," *American Journal of Infection Control*, Vol. 48, No. 4, 2020, pp. 361–367.

Gesser-Edelsburg, A., A. Diamant, R. Hijazi, and G. S. Mesch, "Correcting Misinformation by Health Organizations During Measles Outbreaks: A Controlled Experiment," *PLoS ONE*, Vol. 13, No. 12, 2018, Article e0209505.

Giese, H., H. Neth, M. Moussaïd, C. Betsch, and W. Gaissmaier, "The Echo in Flu-Vaccination Echo Chambers: Selective Attention Trumps Social Influence," *Vaccine*, Vol. 38, No. 8, 2020, pp. 2070–2076.

Glenski, M., T. Weninger, and S. Volkova, "Propagation from Deceptive News Sources: Who Shares, How Much, How Evenly, and How Quickly?" *IEEE Transactions on Computational Social Systems*, Vol. 5, No. 4, 2018, pp. 1071–1082.

Golbeck, J., M. Mauriello, B. Auxier, K. H. Bhanushali, C. Bonk, M. A. Bouzaghrane, C. Buntain, R. Chanduka, P. Cheakalos, J. B. Everett, W. Falak, C. Gieringer, J. Graney, K. M. Hoffman, L. Huth, Z. Ma, M. Jha, M. Khan, V. Kori, E. Lewis, G. Mirano, W. T. Mohn, S. Mussenden, T. M. Nelson, S. McWillie, A. Pant, P. Shetye, R. Shrestha, A. Steinheimer, A. Subramanian, and G. Visnansky, "Fake News Vs Satire: A Data Set and Analysis," *Proceedings of the 10th ACM Conference on Web Science*, Amsterdam, Netherlands, 2018.

Greenberg, J., E. Dube, and M. Driedger, "Vaccine Hesitancy: In Search of the Risk Communication Comfort Zone," *PLoS*, No. 9, March 3, 2017.

Gualda, E., and J. Ruas, "Conspiracy Theories, Credibility and Trust in Information," *Communication and Society-Spain*, Vol. 32, No. 1, 2019, pp. 179–194.

Guess, A. M., M.. Lerner, B. Lyons, J. M. Montgomery, B. Nyhan, J. Reifler, and N. Sircar, "A Digital Media Literacy Intervention Increases Discernment Between Mainstream and False News in the United States and India," *PNAS*, Vol. 117, No. 27, 2020, pp. 15536–15545.

Guidry, J. P. D., K. Carlyle, M. Messner, and Y. Jin, "On Pins and Needles: How Vaccines Are Portrayed on Pinterest," *Vaccine*, Vol. 33, No. 39, 2015, pp. 5051–5056.

Gunaratne, K., E. A. Coomes, and H. Haghbayan, "Temporal Trends in Anti-Vaccine Discourse on Twitter," *Vaccine*, Vol. 37, No. 35, 2019, pp. 4867–4871.

Harvey, A. M., S. Thompson, A. Lac, and F. L. Coolidge, "Fear and Derision: A Quantitative Content Analysis of Provaccine and Anti-Vaccine Internet Memes," *Health Education and Behavior*, Vol. 46, No. 6, 2019, pp. 1012–1023.

He, S., X. Zheng, J. Wang, Z. Chang, Y. Luo, and D. Zeng, "Meme Extraction and Tracing in Crisis Events," Conference on Intelligence and Security Informatics, Tucson, Ariz., 2016, pp. 61–66.

Helmus, Todd C., and Elizabeth Bodine-Baron, *Empowering ISIS Opponents on Twitter*, Santa Monica, Calif.: RAND Corporation, PE-227-RC, 2017. As of March 11, 2021:
https://www.rand.org/pubs/perspectives/PE227.html

Helmus, Todd C., James V. Marrone, and Marek N. Posard, *Russian Propaganda Hits Its Mark: Experimentally Testing the Impact of Russian Propaganda and Counter-Interventions*, Santa Monica, Calif.: RAND Corporation, RR-A704-3, 2020. As of March 12, 2021:
https://www.rand.org/pubs/research_reports/RRA704-3.html

"Historical PowerTrack API," Twitter Developer blog, undated. As of March 21, 2021:
https://developer.twitter.com/en/docs/twitter-api/enterprise/historical-powertrack-api/overview

Hoffman, B. L., E. M. Felter, K. H. Chu, A. Shensa, C. Hermann, T. Wolynn, D. Williams, and B. A. Primack, "It's Not All About Autism: The Emerging Landscape of Anti-Vaccination Sentiment on Facebook," *Vaccine*, Vol. 37 No. 16, 2019, pp. 2216–2223.

Hope, Jonathan, and Michael Witmore, "The Hundredth Psalm to the Tune of "Green Sleeves": Digital Approaches to Shakespeare's Language of Genre,' *Shakespeare Quarterly*, Vol. 61, No. 3, Fall 2010.

Hornsey, M. J., M. Finlayson, G. Chatwood, and C. T. Begeny, "Donald Trump and Vaccination: The Effect of Political Identity, Conspiracist Ideation and Presidential Tweets on Vaccine Hesitancy," *Journal of Experimental Social Psychology*, Vol. 88, 2020, Article 103947.

Hui, P. M., C. Shao, A. Flammini, F. Menczer, and G. L. Ciampaglia, "The Hoaxy Misinformation and Fact-Checking Diffusion Network," *Proceedings of the International AAAI Conference on Web and Social Media*, Palo Alto, Calif., 2018, pp. 528–530.

Hussain, M. N., S. Tokdemir, N. Agarwal, and S. Al-Khateeb, "Analyzing Disinformation and Crowd Manipulation Tactics on YouTube," *Proceedings of the 2018 IEEE/ACM International Conference on Advances in Social Networks Analysis and Mining*, Barcelona, Spain, August 2018, pp. 1092–1095.

Iannelli, L., F. Giglietto, L. Rossi, and E. Zurovac, "Facebook Digital Traces for Survey Research: Assessing the Efficiency and Effectiveness of a Facebook Ad–Based Procedure for Recruiting Online Survey Respondents in Niche and Difficult-to-Reach Populations," *Social Science Computer Review*, Vol. 38, No. 4, 2018.

Jacques, P. J., and C. C. Knox, "Hurricanes and Hegemony: A Qualitative Analysis of Micro-Level Climate Change Denial Discourses," *Environmental Politics*, Vol. 25, No. 5, 2016, pp. 831–852.

Jamison, A. M., D. A. Broniatowski, M. Dredze, Z. Wood-Doughty, D. Khan, and S. C. Quinn, "Vaccine-Related Advertising in the Facebook Ad Archive," *Vaccine*, Vol. 38, No. 3, 2020, pp. 512–520.

Jenkins, M. C., and M. A. Moreno, "Vaccination Discussion Among Parents on Social Media: A Content Analysis of Comments on Parenting Blogs," *Journal of Health Communication*, Vol. 25, No, 3, March 10, 2020, pp. 1–11.

Jolley, Daniel, Rose Meleady, and Karen M. Douglas, "Exposure to Intergroup Conspiracy Theories Promotes Prejudice Which Spreads Across Groups," *British Journal of Psychology*, Vol. 111, No. 1, 2020, pp. 17–35.

Kavanagh, Jennifer, William Marcellino, Jonathan S. Blake, Shawn Smith, Steven Davenport, and Mahlet G. Tebeka, *News in a Digital Age: Comparing the Presentation of News Information over Time and Across Media Platforms*, Santa Monica, Calif.: RAND Corporation, RR-2960-RC, 2019. As of July 31, 2020: https://www.rand.org/pubs/research_reports/RR2960.html

Kavanagh, Jennifer, and Michael D. Rich, *Truth Decay: An Initial Exploration of the Diminishing Role of Facts and Analysis in American Public Life*, Santa Monica, Calif.: RAND Corporation, RR-2314-RC, 2018. As of July 16, 2020: https://www.rand.org/pubs/research_reports/RR2314.html

Kearney, M. D., P. Selvan, M. K. Hauer, A. E. Leader, and P. M. Massey, "Characterizing HPV Vaccine Sentiments and Content on Instagram," *Health Education and Behavior*, Vol. 46, No. 2 (Supplement), 2019, pp. 37–48.

Keelan, J., V. Pavri, R. Balakrishnan, and K. Wilson, "An Analysis of the Human Papilloma Virus Vaccine Debate on MySpace Blogs," *Vaccine*, Vol. 28, No. 6, 2010, pp. 1535–1540.

Klein, C., P. Clutton, and A. G. Dunn, "Pathways to Conspiracy: The Social and Linguistic Precursors of Involvement in Reddit's Conspiracy Theory Forum," *PLoS ONE*, Vol. 14, No. 11, 2019, Article e0225098.

Klein, C., P. Clutton, and V. Polito, "Topic Modeling Reveals Distinct Interests Within an Online Conspiracy Forum," *Frontiers in Psychology*, Vol. 9, 2018, Article 189.

Kou, Y., X. Gui, Y. Chen, and K. H. Pine, "Conspiracy Talk on Social Media: Collective Sensemaking During a Public Health Crisis," *Proceedings of the ACM on Human-Computer Interaction*, 2017, Article 61.

Krishnendhu, V., and L. George, "Drivers and Barriers for Measles Rubella Vaccination Campaign: A Qualitative Study," *Journal of Family Medicine and Primary Care*, Vol. 8, No. 3, 2019, pp. 881–885.

Kudo, T., and J. Richardson, "Sentencepiece: A Simple and Language Independent Subword Tokenizer and Detokenizer for Neural Text Processing," *Proceedings of the 2018 Conference on Empirical Methods in Natural Language Processing: System Demonstrations*, Brussels, Belgium, 2018, pp. 66–71.

Landrum, A. R., A. Olshansky, and O. Richards, "Differential Susceptibility to Misleading Flat Earth Arguments on YouTube," *Media Psychology*, Vol. 24, No. 1, 2019.

Larson, H. J., R. Wilson, S. Hanley, A. Parys, and P. Paterson, "Tracking the Global Spread of Vaccine Sentiments: The Global Response to Japan's Suspension of Its HPV Vaccine Recommendation," *Human Vaccines and Immunotherapeutics*, Vol. 10, No. 9, 2014, pp. 2543–2550.

Lutkenhaus, R. O., J. Jansz, and M. P. A. Bouman, "Mapping the Dutch Vaccination Debate on Twitter: Identifying Communities, Narratives, and Interactions," *Vaccine: X*, Vol. 1, April 2019, Article 100019.

Madden, K., X. Nan, R. Briones, and L. Waks, "Sorting Through Search Results: A Content Analysis of HPV Vaccine Information Online," *Vaccine*, Vol. 30, No. 25, 2012, pp. 3741–3746.

Mahajan, R., W. Romine, M. Miller, and T. Banerjee, "Analyzing Public Outlook Towards Vaccination Using Twitter," 2019 IEEE International Conference on Big Data (Big Data), Los Angeles, Calif., December 9–12, 2019.

Marcellino, William M., "Revisioning Strategic Communication Through Rhetoric and Discourse Analysis," *Joint Force Quarterly*, Vol. 76, No. 1, 2015, pp. 52–57. As of March 30, 2021:
https://ndupress.ndu.edu/Portals/68/Documents/jfq/jfq-76/jfq-75_52-57_Marcellino.pdf

Marcellino, William, Kate Cox, Katerina Galai, Linda Slapakova, Amber Jaycocks, and Ruth Harris, *Human-Machine Detection of Online-Based Malign Information*, Santa Monica, Calif.: RAND Corporation, RR-A519-1, 2020. As of July 20, 2020:
https://www.rand.org/pubs/research_reports/RRA519-1.html

Marcon, A. R., B. Murdoch, and T. Caulfield, "Fake News Portrayals of Stem Cells and Stem Cell Research," *Regenerative Medicine*, Vol. 12, No. 7, 2017, pp. 765–775.

Meleo-Erwin, Z., C. Basch, S. A. MacLean, C. Scheibner, and V. Cadorett, "'To Each His Own': Discussions of Vaccine Decision-Making in Top Parenting Blogs," *Human Vaccines and Immunotherapeutics*, Vol. 13, No. 8, 2017, pp. 1895–1901.

Meylakhs, P., Y. Rykov, O. Koltsova, and S. Koltsov, "An AIDS-Denialist Online Community on a Russian Social Networking Service: Patterns of Interactions with Newcomers and Rhetorical Strategies of Persuasion," *Journal of Medical Internet Research*, Vol. 16, No. 11, 2014, Article e261.

Mitra, T., S. Counts, and J. W. Pennebaker, "Understanding Anti-Vaccination Attitudes in Social Media," *Proceedings of the Tenth International AAAI Conference on Web and Social Media*, Cologne, Germany, 2016.

Mocanu, D., L. Rossi, Q. Zhang, M. Karsai, and W. Quattrociocchi, "Collective Attention in the Age of (Mis)Information," *Computers in Human Behavior*, Vol. 51, 2015, pp. 1198–1204.

Mancosu, M., S. Vassallo, and C. Vezzoni, "Believing in Conspiracy Theories: Evidence from an Exploratory Analysis of Italian Survey Data," *South European Society and Politics*, Vol. 22, No. 3, 2017, pp. 327–344.

Moses, D., "'White Genocide' and the Ethics of Public Analysis," *Journal of Genocide Research*, Vol. 21, No. 2, 2019, pp. 201–213.

Mozafari, M., R. Farahbakhsh, and N. Crespi, *A BERT-Based Transfer Learning Approach for Hate Speech Detection in Online Social Media*, ArXiv.org, Oct. 2019. As of March 15, 2020: http://arxiv.org/abs/1910.12574

Narayan, B., and M. Preljevic, "An Information Behaviour Approach to Conspiracy Theories: Listening in on Voices from Within the Vaccination Debate," *Information Research: An International Electronic Journal*, Vol. 22, No. 1, March 2017, Article 1616.

Nerghes, A., P. Kerkhof, and I. Hellsten, "Early Public Responses to the Zika-Virus on YouTube: Prevalence of and Differences Between Conspiracy Theory and Informational Videos," *Proceedings of the 10th ACM Conference on Web Science*, Amsterdam, Netherlands, 2018.

Nugier, A., F. Limousi, and N. Lydié, "Vaccine Criticism: Presence and Arguments on French-Speaking Websites," *Medecine et Maladies Infectieuses*, Vol. 48, No. 1, 2018, pp. 37–43.

Numerato, D., L. Vochocová, V. Štětka, and A. Macková, "The Vaccination Debate in the "Post-Truth" Era: Social Media as Sites of Multi-Layered Reflexivity," *Sociology of Health and Illness*, Vol. 41, No. S1, 2019, pp. 82–97.

Okoli, C., "A Guide to Conducting a Standalone Systematic Literature Review," *Communications of the Association for Information Systems*, Vol. 37, No. 1, 2015.

Okuhara, T., H. Ishikawa, M. Okada, M. Kato, and T. Kiuchi, "Contents of Japanese Pro- and Anti-HPV Vaccination Websites: A Text Mining Analysis," *Patient Education and Counseling*, Vol. 101, No. 3, 2018, pp. 406–413.

———, "Japanese Anti- Versus Pro-Influenza Vaccination Websites: A Text-Mining Analysis," *Health Promotion International*, Vol. 34, No. 3, 2019, pp. 552–566.

O'Sullivan, Donie, and Konstantin Toropin, "QAnon Fans Spread Fake Claims About Real Fires in Oregon," CNN, September 11, 2020. As of March 11, 2021: https://www.cnn.com/2020/09/11/tech/qanon-oregon-fire-conspiracy-theory/index.html

Penţa, M. A., and A. Băban, "Dangerous Agent or Saviour? HPV Vaccine Representations on Online Discussion Forums in Romania," *International Journal of Behavioral Medicine*, Vol. 21, No. 1, 2014, pp. 20–28.

"Plandemic Part 1," video featuring an interview with Judy Mikovits, via ISE, 2020. As of march 23, 2021: ise.media/video/plandemic-part-1-20.html

Poberezhskaya, M., "Blogging About Climate Change in Russia: Activism, Scepticism and Conspiracies," *Environmental Communication: A Journal of Nature and Culture*, Vol. 12, No. 7, 2018, pp. 942–955.

Porat, T., P. Garaizar, M. Ferrero, H. Jones, M. Ashworth, and M. A. Vadillo, "Content and Source Analysis of Popular Tweets Following a Recent Case of Diphtheria in Spain," *European Journal of Public Health*, Vol. 29, No. 1, 2019, pp. 117–122.

Porreca, A., F. Scozzari, and M. Di Nicola, "Using Text Mining and Sentiment Analysis to Analyse Youtube Italian Videos Concerning Vaccination," *BMC Public Health*, Vol. 20, No. 1, 2020, Article 259.

Pyrhonen, N., and G. Bauvois, "Conspiracies Beyond Fake News: Producing Reinformation on Presidential Elections in the Transnational Hybrid Media System," *Sociological Inquiry*, 2019.

Ringler, H., B. B. Klebanov, and D. Kaufer, "Placing Writing Tasks in Local and Global Contexts: The Case of Argumentative Writing," *Journal of Writing Analytics*, Vol. 2, 2018, pp. 34–77.

Roxburgh, N., D. Guan, K. J. Shin, W. Rand, S. Managi, R. Lovelace, and J. Meng, "Characterising Climate Change Discourse on Social Media During Extreme Weather Events," *Global Environmental Change*, Vol. 54, 2019, pp. 50–60.

Ruder, S., *An Overview of Multi-Task Learning in Deep Neural Networks*, ArXiv.org, June 2017. As of March 15, 2021:
http://arxiv.org/abs/1706.05098

Samantray, A., and P. Pin, "Credibility of Climate Change Denial in Social Media," *Nature*, Vol. 5, No. 1, 2019, Article 127.

Samory, M., and T. Mitra, "Conspiracies Online: User Discussions in a Conspiracy Community Following Dramatic Events," *Proceedings of the International AAAI Conference on Web and Social Media*, Vol. 12, No. 1, 2018a.

———, "'The Government Spies Using Our Webcams': The Language of Conspiracy Theories in Online Discussions," *Proceedings of the ACM on Human-Computer Interaction*, 2018b, Article 152.

Schaeffer, Katherine, "A Look at the Americans who Believe There Is Some Truth to the Conspiracy That COVID-19 Was Planned," *Fact Tank*, Pew Research Center, July 24, 2020, As of September 9, 2020:
https://www.pewresearch.org/fact-tank/2020/07/24/a-look-at-the-americans-who
-believe-there-is-some-truth-to-the-conspiracy-theory-that-covid-19-was-planned/

Schmitt, E. A., and H. Li, "Engaging Truthiness and Obfuscation in a Political Ecology Analysis of a Protest Against the Pengzhou Petroleum Refinery," *Journal of Political Ecology*, Vol. 26, No. 1, 2019, pp. 579–598.

Selepak, A. G., "Exploring Anti-Science Attitudes Among Political and Christian Conservatives Through an Examination of American Universities on Twitter," *Cogent Social Sciences*, Vol. 4, No. 1, 2018, Article 1462134.

Sharma, M., K. Yadav, N. Yadav, and K. C. Ferdinand, "Zika Virus Pandemic— Analysis of Facebook as a Social Media Health Information Platform," *American Journal of Infection Control*, Vol. 45, No. 3, 2017, pp. 301–302.

Smith, N., and T. Graham, "Mapping the Anti-Vaccination Movement on Facebook," *Information Communication and Society*, Vol. 22, No. 9, 2019, pp. 1310–1327.

Sommariva, S., C. Vamos, A. Mantzarlis, L. U. L. Đào, and D. Martinez Tyson, "Spreading the (Fake) News: Exploring Health Messages on Social Media and the Implications for Health Professionals Using a Case Study," *American Journal of Health Education*, Vol. 49, No. 4, 2018, pp. 246–255.

Song, M. Y. J., and A. Gruzd, "Examining Sentiments and Popularity of Pro-and Anti-Vaccination Videos on YouTube," *#SMSociety17: Proceedings of the 8th International Conference on Social Media & Society*, Toronto, Canada, July 2017, pp. 1–8.

Stæhr, A., The Appropriation of Transcultural Flows Among Copenhagen Youth—The Case of Illuminati," *Discourse, Context and Media*, Vol. 4–5, 2018, pp. 101–115.

Steffens, M. S., A. G. Dunn, K. E. Wiley, and J. Leask, "How Organisations Promoting Vaccination Respond to Misinformation on Social Media: A Qualitative Investigation," *BMC Public Health*, Vol. 19, No. 1, 2019, Article 1348.

Straton, N., H. Jang, R. Ng, R. Vatrapu, and R. R. Mukkamala, "Computational Modeling of Stigmatized Behaviour in Pro-Vaccination and Anti-Vaccination Discussions on Social Media," *2019 IEEE International Conference on Bioinformatics and Biomedicine (BIBM)*, 2019, pp. 18–21.

Takaoka, Alicia, "Searching for Community and Safety: Evaluating Common Information Shared in Online Ex-Vaxxer Communities," in G. Meiselwitz, ed., *Social Computing and Social Media: Communication and Social Communities*, Cham, Switzerland: Springer, Lecture Notes in Computer Science, Vol. 11579, 2019, pp. 495–513.

TensorFlow, "bert_en_uncased_L-12_H-768_A-12," software, undated. As of March 15, 2021:
https://tfhub.dev/tensorflow/bert_en_uncased_L-12_H-768_A-12/2

Tingley, D., and G. Wagner, "Solar Geoengineering and the Chemtrails Conspiracy on Social Media," *Nature*, Vol. 3, No. 1, 2017, Article 14.

Tomeny, T. S., C. J. Vargo, and S. El-Toukhy, "Geographic and Demographic Correlates of Autism-Related Anti-Vaccine Beliefs on Twitter, 2009–15," *Social Science and Medicine*, No. 191, 2017, pp. 168–175.

Uscinski, J. E., A. M. Enders, C. Klofstad, M. Seelig, J. Funchion, C. Everett, S. Wuchty, K. Premaratne, and M. Murthi, "Why Do People Believe COVID-19 Conspiracy Theories?" *Harvard Kennedy School Misinformation Review*, Vol. 1, No. 3, 2020.

van der Linden, S., E. Maibach, J. Cook, A. Leiserowitz, and S. Lewandowsky, "Inoculating Against Misinformation," *Science*, Vol. 358, 2017, pp. 1141–1142.

Vijaykumar, S., G. Nowak, I. Himelboim, and Y. Jin, "Virtual Zika Transmission After the First U.S. Case: Who Said What and How It Spread on Twitter," *American Journal of Infection Control*, Vol. 46, No. 5, 2018, pp. 549–557.

Vraga, E. K., and L. Bode, "Using Expert Sources to Correct Health Misinformation in Social Media," *Science Communication*, Vol. 29, No. 5, 2017, pp. 621–645.

Walker, S. , and H. C. Maltezou, "Vaccine-Preventable Diseases in Europe: Where Do We Stand?" *Expert Review of Vaccines*, Vol. 13, No. 8, 2014, pp. 979–987.

Walter, N., J. J. Brooks, C. J. Saucier, and S. Suresh, "Evaluating the Impact of Attempts to Correct Health Misinformation on Social Media: A Meta-Analysis," *Health Communication*, 2020, pp. 1–9.

Wong, L. P., P. F. Wong, and S. AbuBakar, "Vaccine Hesitancy and the Resurgence of Vaccine Preventable Diseases: The Way Forward for Malaysia, a Southeast Asian Country," *Human Vaccines and Immunotherapeutics*, Vol. 16, No. 7, 2020, pp. 1511–1520.

Wood, M. J.,"Propagating and Debunking Conspiracy Theories on Twitter During the 2015–2016 Zika Virus Outbreak," *Cyberpsychology, Behavior, and Social Networking*, Vol. 21, No. 8, 2018, pp. 485–490.

Xuan, L., and A. MacDonald, "Examining Psychosis in Social Media: The Targeted Individuals Movement and the Potential of Pathological Echo-Chambers," *Schizophrenia Bulletin*, Vol. 45, No. 2, 2019, pp. S250–S251.

Yuan, X., R. J. Schuchard, and A. T. Crooks, "Examining Emergent Communities and Social Bots Within the Polarized Online Vaccination Debate in Twitter," *Social Media and Society*, Vol. 5, No. 3, 2019.

Zollo, F., A. Bessi, M. Del Vicario, A. Scala, G. Caldarelli, L. Shekhtman, S. Havlin, and W. Quattrociocchi, "Debunking in a World of Tribes," *PLoS ONE*, Vol. 12, No. 7, 2017, Article e0181821.

Zollo, F., P. K. Novak, M. Del Vicario, A. Bessi, I. Mozetič, A. Scala, G. Caldarelli, W. Quattrociocchi, and T. Preis, "Emotional Dynamics in the Age of Misinformation," *PLoS ONE*, Vol. 10, No. 9, 2015, Article e138740.